Holt McDougal Mathematics

Course 2
Chapter 6 Resource Book

ISBN 13: 978-0-55-400737-3
ISBN 10: 0-55-400737-1

4 5 6 7 8 9   1417   13 12 11 10

4500256713

Contents

Section 6A Family Letter 1
Section 6A At-Home Practice 3
Section 6A Family Fun 4
6-1 Practice A, B, C 5
6-1 Review for Mastery........................ 8
6-1 Challenge.................................. 9
6-1 Problem Solving 10
6-1 Reading Strategies...........................11
6-1 Puzzles, Twisters & Teasers 12
6-2 Practice A, B, C..................... 13
6-2 Review for Mastery........................ 16
6-2 Challenge.................................. 17
6-2 Problem Solving 18
6-2 Reading Strategies........................ 19
6-2 Puzzles, Twisters & Teasers 20
6-3 Practice A, B, C..................... 21
6-3 Review for Mastery........................ 24
6-3 Challenge.................................. 26
6-3 Problem Solving 27
6-3 Reading Strategies........................ 28
6-3 Puzzles, Twisters & Teasers 29
6-4 Practice A, B, C..................... 30
6-4 Review for Mastery........................ 33
6-4 Challenge.................................. 34
6-4 Problem Solving 35
6-4 Reading Strategies........................ 36
6-4 Puzzles, Twisters & Teasers 37
6-5 Practice A, B, C..................... 38
6-5 Review for Mastery........................ 41
6-5 Challenge.................................. 42
6-5 Problem Solving 43
6-5 Reading Strategies........................ 44
6-5 Puzzles, Twisters & Teasers 46
Section 6B Family Letter........................47
Section 6B At-Home Practice 48
Section 6B Family Fun...........................49
6-6 Practice A, B, C..................... 50
6-6 Review for Mastery........................ 53
6-6 Challenge.................................. 54
6-6 Problem Solving 55
6-6 Reading Strategies........................ 56
6-6 Puzzles, Twisters & Teasers 57
6-7 Practice A, B, C..................... 58
6-7 Review for Mastery........................ 61
6-7 Challenge.................................. 62
6-7 Problem Solving 63
6-7 Reading Strategies........................ 64
6-7 Puzzles, Twisters & Teasers 65
Answers 66

Holt McDougal Mathematics

Description of Contents

Family Involvement Pages

The Chapter Resource Book includes a set of family involvement pages for each section. The family involvement pages consist of the following items:

- **Family Letter**, a two-page letter describing the math that the student will study in each section. A list of vocabulary words from the lessons is included, and some worked-out examples are provided.

- **At-Home Practice**, a one-page worksheet with problems drawn from the content described in the Family Letter. Answers are provided on the page so that the student and his or her family can check this work at home.

- **Family Fun**, a one-page activity sheet that the student and his or her family can work on together.

Practice A, B, and C

There are three practice worksheets for every lesson. All of these reinforce the content of the lesson. Practice B is shown in the Teacher's Edition and is appropriate for the on-level student. It is also available as a workbook (Homework & Practice Workbook)

Practice A is easier than Practice B but still practices the content of the lesson. Practice C is more challenging than Practice B.

Review for Mastery

The Review for Mastery worksheet (one per lesson) provides an alternate way to teach or review the main concepts of the lesson. This worksheet is one or two pages long and is shown in the Teacher's Edition.

Challenge

The Challenge worksheet (one per lesson) enhances critical thinking skills and extends the lesson. This worksheet is shown in the Teacher's Edition.

Problem Solving

The Problem Solving worksheet (one per lesson) provides practice in problem solving and opportunities for real-world applications and for interdisciplinary connections. There are both multiple choice and short response problems. This worksheet is shown in the Teacher's Edition.

Reading Strategies

The Reading Strategies worksheet (one per lesson) provides tools to help the student master math vocabulary or symbols.

Puzzles, Twisters & Teasers

The Puzzles, Twisters & Teasers worksheet (one per lesson) provides fun practice while reinforcing the content of the lesson.

     Holt McDougal Mathematics

Family Letter

6A Proportions and Percents

Dear Family,

The student will be learning about percents, decimals, and fractions. Ask the student, "*What is a percent?*" He or she might tell you that a **percent** is a ratio of a number to 100 and is written using the % symbol. Since a percent represents a part of a whole, it can be modeled much like a fraction or decimal, as shown. 50% or 50 out of 100 squares are shaded.

$$50\% \quad = \quad \frac{50}{100} \quad = \quad 0.50$$

Percent   Fraction   Decimal

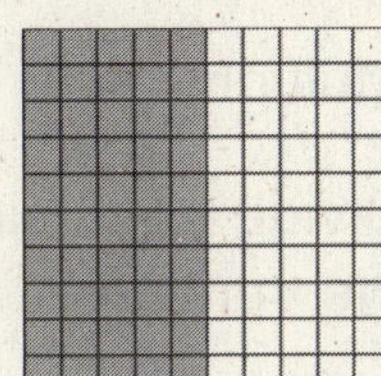

Write 35% as a fraction in simplest form.

$35\% = \dfrac{35}{100}$ Write the percent as a fraction with a denominator of 100.

$\dfrac{35}{100} = \dfrac{7}{20}$ Simplify.

Write 62% as a decimal.

$62\% = 62.0\%$

$62.0\% = 0.62$ Move the decimal point two places to the left.

Write $\dfrac{5}{8}$ as a percent.

$\dfrac{5}{8} = 5 \div 8 = 0.625$ Use division to write the fraction as a decimal.

$0.625 = 62.5\%$ Write the decimal as a percent.

A decimal can be changed to a percent by moving the decimal point two places to the right and adding the percent sign.
$0.1 = 0.10 = 10\%$

Learning to estimate percents can help the student determine whether an answer is reasonable.

Use a fraction to estimate 27% of 57.

$27\% \text{ of } 57 \approx \dfrac{1}{4} \text{ of } 57.$ 27% is about 25%, and $25\% = \dfrac{1}{4}$.

$\approx \dfrac{1}{4} \cdot 60$ Round 57 to a compatible number and multiply.

≈ 15 27% of 57 is about 15.

Vocabulary

These are the math words we are learning:

percent a ratio of a number to 100

Holt McDougal Mathematics

The student will use proportions and decimal equivalents to find the percent of a number.

Find 23% of 60.

$$\frac{23}{100} = \frac{n}{60}$$ Write a proportion.

$23 \cdot 60 = 100 \cdot n$ Set the cross products equal.

$1{,}380 = 100n$ Multiply.

$$\frac{1{,}380}{100} = \frac{100n}{100}$$ Divide each side by 100 to isolate the variable.

$13.8 = n$

13.8 is 23% of 60.

The student will solve problems containing percents by using either proportions or equations.

What percent of 60 is 18?

$$\frac{n}{100} = \frac{18}{60}$$ Write a proportion.

$n \cdot 60 = 100 \cdot 18$ Set the cross products equal.

$60n = 1{,}800$ Multiply.

$n = 30$ Divide each side by 60.

30% of 60 is 18.

72 is 20% of what number?

$72 = 20\% \cdot n$ Write an equation.

$72 = 0.2 \cdot n$ Write 20% as a decimal.

$360 = n$ Divide each side by 0.2.

72 is 20% of 360.

Work through some of the practice problems with the student. Your involvement will encourage his or her interest in mathematics.

Sincerely,

At-Home Practice

6A Proportions and Percents

Write each percent as a fraction in simplest form.

1. 25%

2. 49%

3. 55%

Write each percent as a decimal.

4. 29%

5. 1%

6. 36%

Write each decimal or fraction as a percent.

7. 0.26

8. $\dfrac{6}{10}$

9. $\dfrac{10}{25}$

Use a fraction to estimate the percent of each number.

10. 18% of 151

11. 34% of 90

12. 52% of 269

Find the percent of each number.

13. 26% of 20

14. 12% of 75

15. 25% of 95

Solve.

16. 60 is 120% of what number?

17. 48 is what percent of 20?

18. 13 is what percent of 65?

19. 3 is 15% of what number?

20. 310 is 25% of what number?

21. 6 is what percent of 12?

Answers: 1. $\dfrac{1}{4}$ **2.** $\dfrac{49}{100}$ **3.** $\dfrac{11}{20}$ **4.** 0.29 **5.** 0.01 **6.** 0.36 **7.** 26% **8.** 60% **9.** 40% **10.** ≈ 30 **11.** ≈ 30 **12.** ≈ 135

13. 5.2 **14.** 9 **15.** 23.75 **16.** 50 **17.** 240% **18.** 20% **19.** 20 **20.** 1,240 **21.** 50%

Family Fun

Matching Game with Percents

Materials
24 fraction/decimal/percent cards

Directions
• Cut out the cards. Shuffle the cards and place them face down.

• Players take turns turning over 2 cards.

• If the cards are equivalent, the player keeps them and gets another turn.

• If the cards are not equivalent, the cards are turned back over in the same place.

• The game is over when all the cards have been matched.

• The player with the most matched cards is the winner.

10%	50%	40%	30%	25%	75%
100%	1%	90%	120%	2%	70%
1.2	0.25	1	0.4	$\frac{3}{10}$	$\frac{3}{4}$
$\frac{1}{100}$	$\frac{9}{10}$	$\frac{14}{20}$	$\frac{1}{10}$	$\frac{1}{2}$	$\frac{1}{50}$

Holt McDougal Mathematics

Practice A
LESSON 6-1

Percents

Write the fraction of the grid that is shaded. Then write the percent.

1. 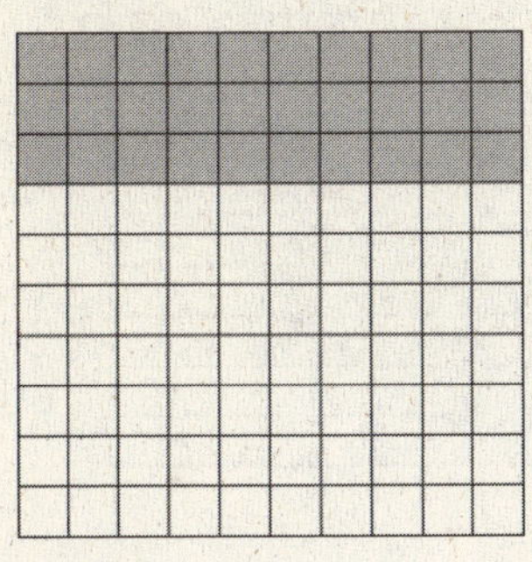

2. 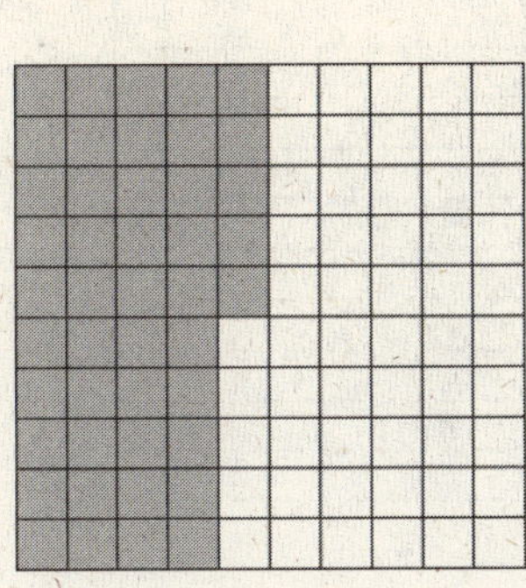

3. 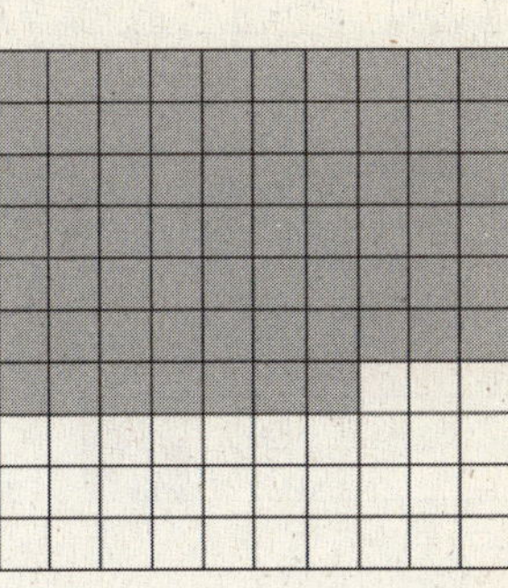

$\dfrac{30}{100} =$ _____ % $\dfrac{}{100} =$ _____ % $\dfrac{}{100} =$ _____ %

Write each percent as a fraction in simplest form.

4. 20%   5. 32%   6. 90%   7. 43%

_________   _________   _________   _________

8. 48%   9. 58%   10. 71%   11. 80%

_________   _________   _________   _________

12. 28%   13. 85%   14. 42%   15. 52%

_________   _________   _________   _________

Write each percent as a decimal.

16. 19%   17. 43%   18. 7%   19. 62%

_________   _________   _________   _________

20. 23%   21. 36%   22. 59%   23. 3%

_________   _________   _________   _________

24. 72.5%   25. 9.8%   26. 7.04%   27. 12.49%

_________   _________   _________   _________

Holt McDougal Mathematics

LESSON 6-1

Practice B

Percents

Write the percent modeled by each grid.

1. 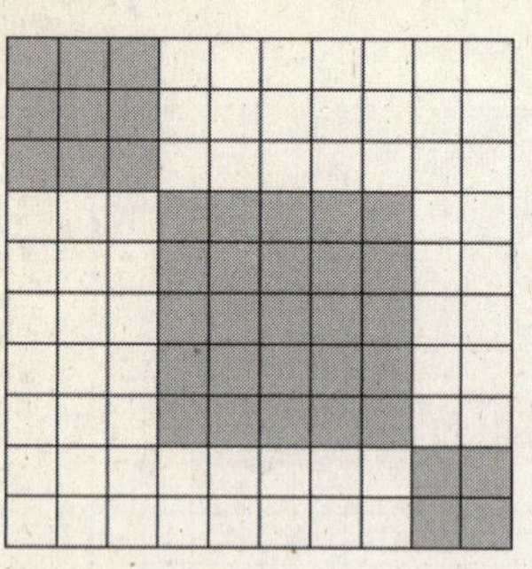 2. 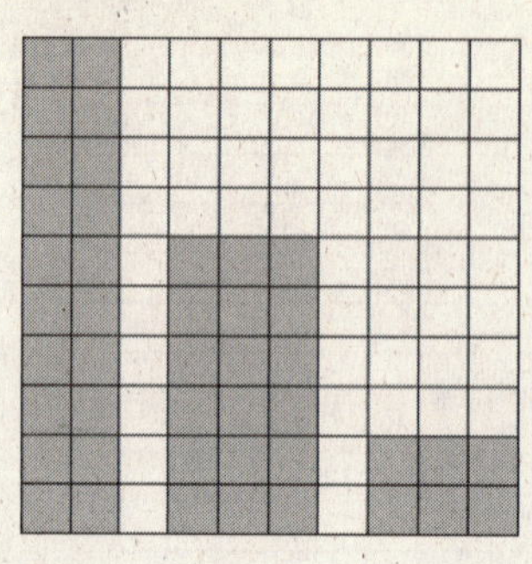 3. 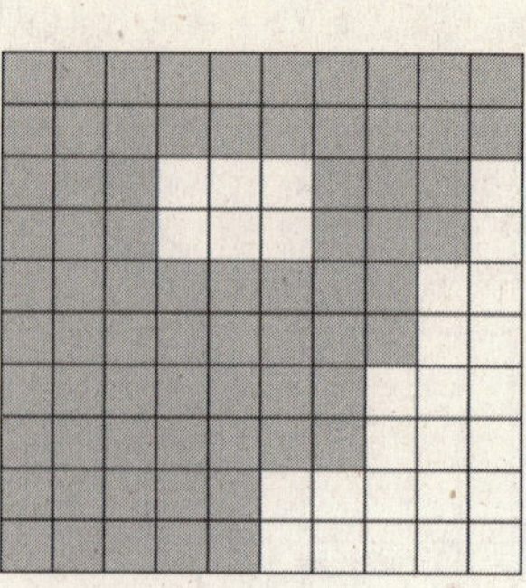

_____________________   _____________________   _____________________

Write each percent as a fraction in simplest form.

4. 16% _____________   5. 49% _____________   6. 20% _____________   7. 15% _____________

8. 18% _____________   9. 60% _____________   10. 35% _____________   11. 46% _____________

12. 86% _____________   13. 79% _____________   14. 56% _____________   15. 45% _____________

Write each percent as a decimal.

16. 33% _____________   17. 57% _____________   18. 46% _____________   19. 6% _____________

20. 4.7% _____________   21. 13.2% _____________   22. 75.8% _____________   23. 4% _____________

24. 1.16% _____________   25. 27.05% _____________   26. 93.01% _____________   27. 7.9% _____________

Holt McDougal Mathematics

LESSON 6-1 — Practice C
Percents

Write the percent modeled by each grid.

1. 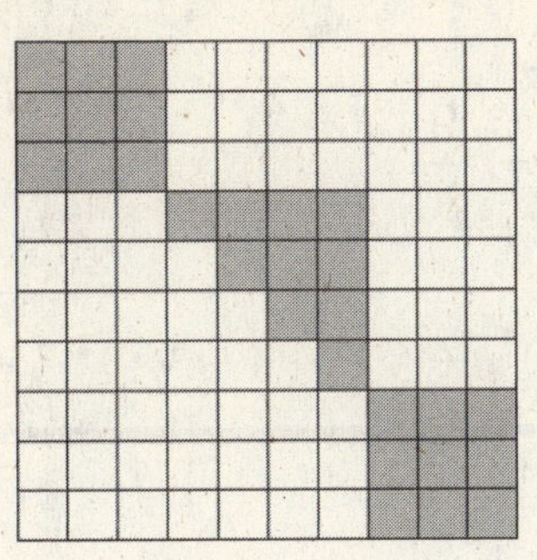

2. 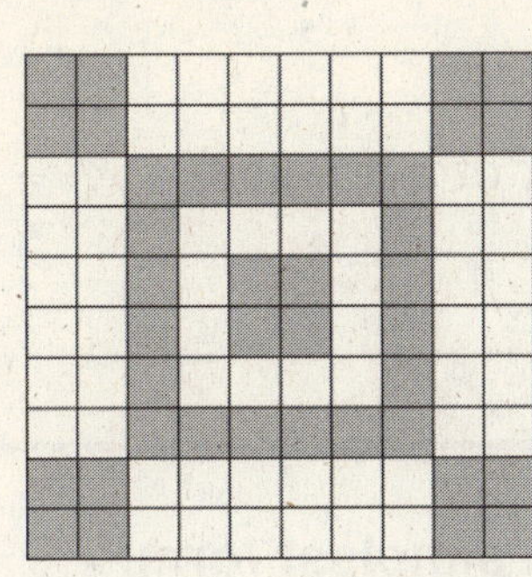

3. 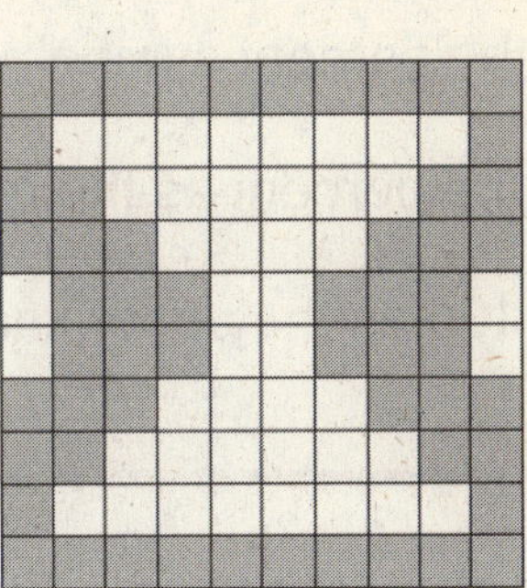

______________________   ______________________   ______________________

Write each percent as a fraction in simplest form and as a decimal.

4. 29%   5. 87.5%   6. 7.5%   7. 64%

______________   ______________   ______________   ______________

8. 62.5%   9. 34%   10. 22.5%   11. 12.5%

______________   ______________   ______________   ______________

12. 3.6%   13. 5.9%   14. 7.2%   15. 20.5%

______________   ______________   ______________   ______________

Compare. Write <, >, or =.

16. $\dfrac{19}{100}$ ▪ 9%   17. $\dfrac{44}{100}$ ▪ 44%   18. $\dfrac{38}{100}$ ▪ 39%

______________   ______________   ______________

19. $\dfrac{12}{25}$ ▪ 40%   20. $\dfrac{19}{20}$ ▪ 96%   21. $\dfrac{27}{50}$ ▪ 52%

______________   ______________   ______________

22. $\dfrac{7}{40}$ ▪ 17.5%   23. $\dfrac{5}{12}$ ▪ 41%   24. $\dfrac{1}{16}$ ▪ 6.5%

______________   ______________   ______________

Holt McDougal Mathematics

LESSON 6-1	**Review for Mastery**
	Percents

To change a percent to a fraction:

• drop the percent symbol;

• write the percent as the numerator of a fraction; $45\% = \dfrac{45}{100} = \dfrac{9}{20}$

• write 100 as the denominator;

• simplify.

Write each percent as a fraction in simplest form.

1. $38\% = \dfrac{}{100} = $ _________

2. $20\% = \dfrac{}{100} = $ _________

3. $70\% = \dfrac{}{100} = $ _________

4. $16\% = \dfrac{}{100} = $ _________

5. $36\% = \dfrac{}{100} = $ _________

6. $8\% = \dfrac{}{100} = $ _________

7. $15\% = $ _________

8. $53\% = $ _________

9. $24\% = $ _________

10. $17\% = $ _________

To change a percent to a decimal:

• drop the percent symbol; $45\% = .45. = 0.45$

• move the decimal point two places to the left. $7\% = .07. = 0.07$

Write each percent as a decimal.

11. 58%

12. 93%

13. 15%

14. 9%

____

15. 26%

16. 2%

17. 80%

18. 1%

____

19. 23.5%

20. 9.6%

21. 40.7%

22. 7.03%

____

                                                      Holt McDougal Mathematics

LESSON 6-1 Challenge

Complementary Events

The probability of an event can be expressed as a percent, a fraction, or a decimal.

When two events are the only two events that can occur, they are called **complementary events**. If two events are complementary, the sum of their probabilities is one. For example, suppose the probability of rain is 60%, or 0.6: $P(rain) = 0.6$. That means the probability of it not raining is 40%, or 0.4: $P(not\ rain) = 0.4$. So, $P(rain\ or\ not\ rain) = 0.6 + 0.4 = 1$.

Draw a line to connect the probability for each Event A to the probability for an Event B that could be a complementary event.

Event A	**Event B**	
$P(A) = 45\%$	$P(B) = 0.29$	**F**
	$P(B) = \dfrac{5}{8}$	**T**
$P(A) = 70\%$	$P(B) = 0.35$	**L**
	$P(B) = 0.175$	**N**
$P(A) = 42\%$	$P(B) = \dfrac{13}{25}$	**E**
	$P(B) = \dfrac{11}{20}$	**A**
$P(A) = 17.5\%$	$P(B) = \dfrac{12}{25}$	**I**
	$P(B) = 0.42$	**S**
$P(A) = 65\%$	$P(B) = \dfrac{3}{10}$	**R**
	$P(B) = \dfrac{9}{20}$	**H**
$P(A) = 37.5\%$	$P(B) = \dfrac{33}{40}$	**U**
$P(A) = 48\%$	$P(B) = \dfrac{29}{50}$	**M**

Look for the probabilities that do <u>not</u> have lines drawn to them. Use the letters next to those probabilities to write the answer to the riddle. You may use each letter as many times as you wish.

What language do sharks speak?

____ ____   ____ ____ ____ ____ ____

     Holt McDougal Mathematics

LESSON 6-1 — Problem Solving
Percents

Write the correct answer.

1. In 2003, 68% of the T.V. set owners in the United States had cable television. Write this percent as a fraction in simplest form and as a decimal.

2. In 2004, 27% of Internet users were in the United States. What percent of Internet users were in countries other than the United States?

3. In a survey, 46% of men said they spend fewer than 5 hours shopping for gifts for the holidays. Write this percent as a fraction in simplest form and as a decimal.

4. In a survey, 59% of a group of people aged 18–29 said that they do not have enough time to do what they want. What percent of those surveyed feel that they do have enough time do what they want? Write your answer as a percent and as a decimal.

Choose the letter for the best answer.

The table shows the percent of adults who participated in selected leisure activities two or more times per week.

5. Express the percent of adults who dined out two or more times per week as a fraction.

   A $\dfrac{1}{100}$ C $\dfrac{1}{10}$

   B $\dfrac{1}{50}$ D $\dfrac{1}{5}$

6. Express the percent of adults who did crossword puzzles two or more times per week as a decimal.

   F 0.07   H 0.5

   G 0.05   J 0.7

Adult Participation in Selected Leisure Activities in 2003

Activity	Percent
Crossword puzzles	7%
Dining out	10%
Reading books	21%
Surfing the net	18%
Video games	5%

7. What fraction of adults played video games fewer than two times per week?

   A $\dfrac{1}{20}$ C $\dfrac{19}{50}$

   B $\dfrac{1}{50}$ D $\dfrac{19}{20}$

8. Which decimal represents the percent of adults who surfed the net fewer than two times per week?

   F 0.018   H 0.18

   G 0.05   J 0.82

Holt McDougal Mathematics

LESSON 6-1	**Reading Strategies**

Reading Strategies
Multiple Representations

Percent means "per hundred." You can use a hundred grid to picture percents.

The shaded part of the grid
is 20% → Read: "20 percent."

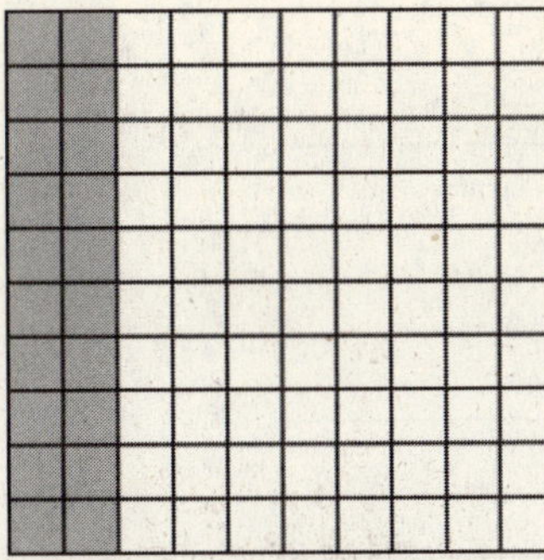

Percents can be written as fractions and as decimals.

20% means "20 per hundred" or $\frac{20}{100}$. The fraction can be written

in simplest form. → $\frac{1}{5}$

$\frac{20}{100}$ is read → "20 hundredths" and can be written → 0.20.

The shaded part of the hundred grid can be represented in these ways:

20% → $\frac{20}{100}$ → $\frac{1}{5}$ → 20 hundredths → 0.20

Answer the following questions.

1. What does the word "percent" mean?_______________________________________

2. What two other ways can you write percents?_______________________________

3. Shade 40% of the hundred grid.

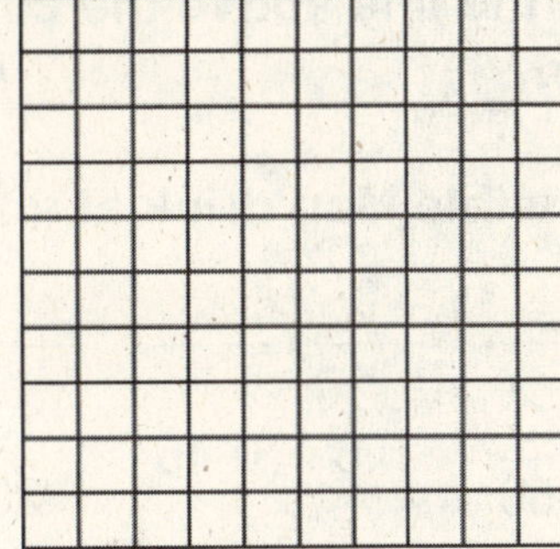

4. Write the shaded part as a decimal. _____________________________________

5. Write the shaded part as a fraction. _______________ or _______________

6. If $\frac{1}{2}$ of the grid were shaded, what percent would that be?_______________

　　　　　　　　　　　　　　　　Holt McDougal Mathematics

LESSON 6-1

Puzzles, Twisters & Teasers
Now You See It...

Write each percent as a fraction in simplest form.

1. 65%   O   _______________

2. 20%   E   _______________

3. 72%   T   _______________

4. 63%   R   _______________

5. 6%   V   _______________

6. 24%   K   _______________

Write each percent as a decimal.

7. 3.62%   L   _______________

8. 60%   A   _______________

9. 52%   I   _______________

10. 6.3%   M   _______________

11. 45%   P   _______________

12. 36.2% D   _______________

Write the letter on the line above the correct answer to solve the riddle.

What does the Invisible Man drink at snack times?

He drinks ____ ____ ____ ____ ____ ____ ____ ____ ____ ____

$\dfrac{1}{5}$ $\dfrac{3}{50}$ 0.6 0.45 $\dfrac{13}{20}$ $\dfrac{63}{100}$ 0.6 $\dfrac{18}{25}$ $\dfrac{1}{5}$ 0.362

____ ____ ____ ____

0.063 0.52 0.0362 $\dfrac{6}{25}$

Holt McDougal Mathematics

LESSON 6-2 — Practice A
Fractions, Decimals, and Percents

Compare. Write <, >, or =.

1. 0.16 ☐ $\dfrac{1}{10}$

2. $\dfrac{3}{5}$ ☐ 60%

3. $\dfrac{3}{10}$ ☐ 0.03

4. $\dfrac{12}{50}$ ☐ $\dfrac{4}{10}$

5. 34% ☐ $\dfrac{3}{4}$

6. 32% ☐ 0.23

Write each decimal as a percent.

7. 0.34

8. 0.08

9. 0.19

10. 0.64

11. 0.561

12. 0.049

13. 0.105

14. 0.9

Write each fraction as a percent.

15. $\dfrac{1}{4}$

16. $\dfrac{7}{50}$

17. $\dfrac{15}{16}$

18. $\dfrac{9}{20}$

19. $\dfrac{7}{8}$

20. $\dfrac{9}{40}$

21. $\dfrac{11}{50}$

22. $\dfrac{4}{5}$

Decide whether to use pencil and paper, mental math, or a calculator. Then solve.

23. In a survey, 25 students were asked whether they prefer orange juice or grapefruit juice. Seventeen students said they prefer orange juice. What percent of the students surveyed said they prefer orange juice?

Holt McDougal Mathematics

LESSON 6-2

Practice B

Fractions, Decimals, and Percents

Write each decimal as a percent.

1. 0.17

2. 0.56

3. 0.04

4. 0.7

5. 0.025

6. 0.803

7. 0.3

8. 0.072

Write each fraction as a percent.

9. $\dfrac{13}{40}$

10. $\dfrac{3}{5}$

11. $\dfrac{3}{20}$

12. $\dfrac{5}{12}$

13. $\dfrac{5}{16}$

14. $\dfrac{3}{80}$

15. $\dfrac{5}{6}$

16. $\dfrac{19}{25}$

Order the numbers from least to greatest.

17. $0.3, \dfrac{19}{40}, 22\%$

18. $11\%, \dfrac{1}{11}, \dfrac{2}{19}$

19. $\dfrac{5}{8}, 0.675, 5.8\%$

20. $-0.35, 0.051, -21\%$

21. $\dfrac{1}{9}, 0.351, 27\%$

22. $0.08, -\dfrac{4}{5}, 0.8$

Decide whether pencil and paper, mental math, or a calculator is most useful when solving the following problem. Then solve.

23. The police use a speed gun to monitor one part of a highway. During one hour, 6 out of 25 cars were traveling above the speed limit. What percent of the cars were traveling above the speed limit?

Holt McDougal Mathematics

LESSON 6-2 — Practice C
Fractions, Decimals, and Percents

Write each decimal as a percent.

1. 0.04   2. 0.975   3. 0.031   4. 0.409

5. 0.378   6. 0.4   7. 0.524   8. 0.067

Write each fraction as a percent.

9. $\dfrac{13}{80}$ 10. $\dfrac{7}{9}$ 11. $\dfrac{17}{20}$ 12. $\dfrac{5}{8}$

13. $\dfrac{39}{40}$ 14. $\dfrac{4}{11}$ 15. $\dfrac{11}{25}$ 16. $\dfrac{9}{50}$

Order the numbers from least to greatest.

17. -12%, $\dfrac{3}{25}$, -0.41 18. $\dfrac{3}{7}$, $\dfrac{4}{9}$, 43% 19. $-\dfrac{11}{12}$, 94.5%, 0.98

20. $\dfrac{5}{6}$, $\dfrac{17}{20}$, 83% 21. $\dfrac{9}{2}$, 0.41, 43.5% 22. 0.55%, 48%, $\dfrac{13}{24}$

Decide whether pencil and paper, mental math, or a calculator is most useful when solving the following problem. Then solve.

23. In a survey, 80 students were asked to name their favorite subject. Thirty students said that English was their favorite. What percent of the students surveyed said that English was their favorite subject?

____

____

    Holt McDougal Mathematics

LESSON 6-2

Review for Mastery

Fractions, Decimals, and Percents

To change a decimal to a percent:
- move the decimal point two places to the right; $0.07 = .07. = 7\%$
- write the % symbol after the number.

Write each decimal as a percent.

1. 0.34

2. 0.06

3. 0.93

4. 0.57

5. 0.8

6. 0.734

7. 0.082

8. 0.225

9. 0.604

10. 0.09

11. 0.518

12. 0.039

To change a fraction to a percent:
- Find an equivalent fraction with a denominator of 100.
- Use the numerator of the equivalent fraction as the percent.

$$\frac{8}{25} = \frac{x}{100}$$

$$\frac{8 \cdot 4}{25 \cdot 4} = \frac{32}{100}$$

$$\frac{8}{25} = \frac{32}{100} = 32\%$$

Think: $100 \div 25 = 4$. So, multiply the numerator and denominator by 4.

Write each fraction as a percent.

13. $\dfrac{3}{10}$

14. $\dfrac{2}{50}$

15. $\dfrac{7}{20}$

16. $\dfrac{1}{5}$

17. $\dfrac{1}{8}$

18. $\dfrac{3}{25}$

19. $\dfrac{3}{4}$

20. $\dfrac{23}{40}$

21. $\dfrac{11}{20}$

22. $\dfrac{43}{50}$

23. $\dfrac{24}{25}$

24. $\dfrac{7}{8}$

Holt McDougal Mathematics

LESSON
6-2

Challenge

Climbing a Percent Pyramid

Begin by changing each ratio in the
pyramid to a percent. Then climb to the
top by comparing the percents as you go.

The percents must increase as you climb
to the top. Each brick on your path must
touch the previous brick. Circle the ratios
in the path that leads to the top.

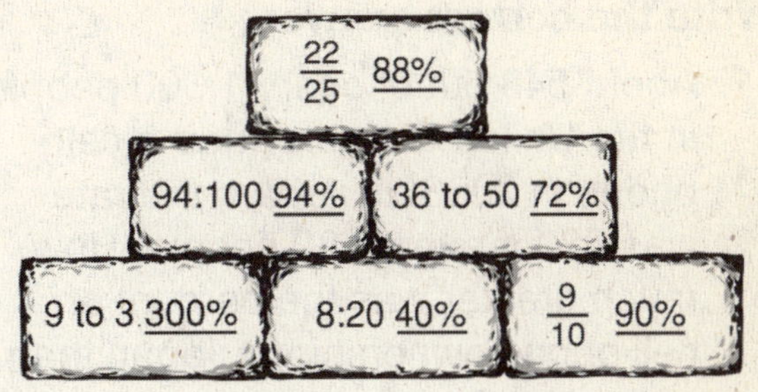

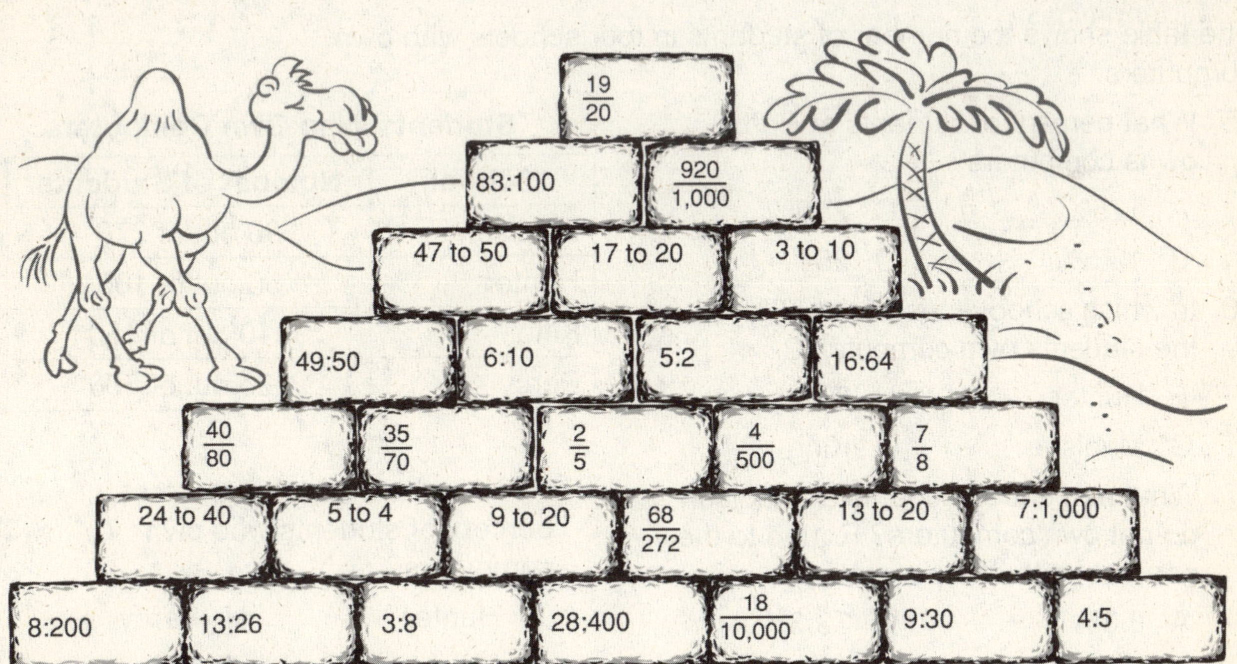

Holt McDougal Mathematics

LESSON 6-2 · Problem Solving
Fractions, Decimals, and Percents

Write the correct answer.

1. About 543 out of every 1,000 people in the United States owned a cell phone in 2003. In Japan, the rate was 68 for every 100 people. How much greater was the percent of cell phone ownership in Japan than in the U.S.?

2. In 2002, the adult population of the United States was about 206 million. About 113 million people participated in an exercise program. To the nearest percent, what percent of the adult population participated in an exercise program?

3. When asked about their favorite Thanksgiving leftover, $\frac{1}{20}$ of the people said vegetables and $\frac{7}{100}$ said mashed potatoes. Which food was more popular and by what percent?

4. In a survey, 80 people were asked whether they thought the speed limit for interstate highways should be raised. Twenty-five people said the speed limit should be raised. What percent of people did not think that the speed limit should be raised?

Choose the letter for the best answer.

The table shows the number of students in four schools who own computers.

5. What percent of students at Percy owns computers?

   A 125% C 250%

   B 12.5% D 25%

6. In which school does about 73% of the students own computers?

   F Hunter H Percy

   G Madison J King

7. What percent of students at Madison do not own computers? Round to the nearest tenth of a percent.

   A 3.3% C 33.3%

   B 9.9% D 66.7%

Students Who Own Computers

School	Number of Students
Madison	90 out of 270
Hunter	56 out of 100
King	110 out of 150
Percy	125 out of 500

8. Which school has the greatest percent of students who own computers?

   F Hunter H Percy

   G Madison J King

Holt McDougal Mathematics

Reading Strategies

LESSON 6-2

Use a Diagram

You can use a diagram to understand the relationships among fractions, decimals, and percents.

Write 120 out of 400 as a fraction in tenths, as a decimal, and as a percent.

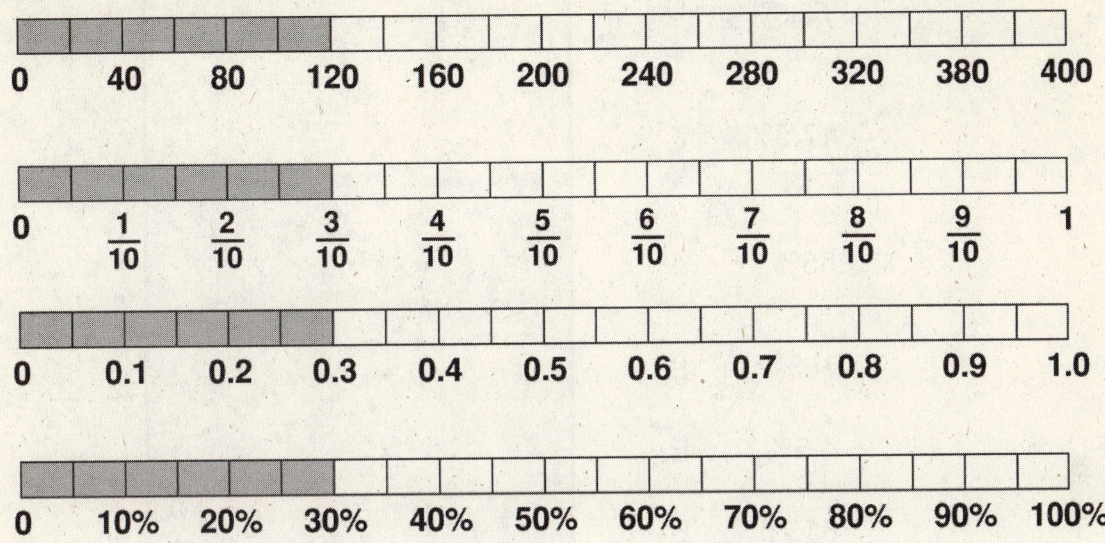

The diagram shows that 120 out of 400 can be written as $\frac{3}{10}$, 0.3, or 30%.

Use the diagram to answer each question.

1. What fraction in tenths can you write for 280 out of 400?

____

2. What percent can you write for 280 out 400?

____

3. What decimal can you write for 180 out of 400?

____

4. What percent can you write for 180 out 400?

____

5. How could you change or add to the diagram above so that you could use it to find the percent that is equal to 110 out of 200?

____

____

       Holt McDougal Mathematics

LESSON
6-2 **Puzzles, Twisters & Teasers**

Pick a Percent

Draw a line from each fraction or decimal to the equivalent percent.

1.	$\dfrac{3}{4}$	55.$\overline{5}$%	U
2.	$\dfrac{5}{9}$	17.5%	I
3.	$\dfrac{13}{16}$	55%	A
4.	0.84	75%	S
5.	0.55	4.5%	E
6.	0.075	81.25%	B
7.	$\dfrac{7}{40}$	82.5%	S
8.	$\dfrac{9}{20}$	7.5%	R
9.	0.045	84%	M
10.	0.825	45%	N

**Write the letter on the line above the matching problem number
to solve the riddle.**

What kind of sandwiches do sailors eat?

___ ___ ___ ___ ___ ___ ___ ___ ___ ___

  1.   2.   3.   4.   5.   6.   7.   8.   9.   10.

 Holt McDougal Mathematics

Name ___________________________________ Date _________________ Class _______________

Practice A

Estimating with Percents

**Which fraction is the best estimate for the given percent?
Choose the letter of the best answer.**

1. 24% A $\dfrac{1}{2}$ B $\dfrac{1}{3}$ C $\dfrac{1}{4}$ D $\dfrac{1}{5}$

2. 73% F $\dfrac{4}{5}$ G $\dfrac{3}{4}$ H $\dfrac{2}{3}$ J $\dfrac{1}{2}$

3. 51% A $\dfrac{1}{2}$ B $\dfrac{2}{3}$ C $\dfrac{3}{4}$ D $\dfrac{4}{5}$

4. 32% F $\dfrac{1}{5}$ G $\dfrac{1}{4}$ H $\dfrac{1}{3}$ J $\dfrac{1}{2}$

5. 39% A $\dfrac{1}{5}$ B $\dfrac{2}{5}$ C $\dfrac{3}{5}$ D $\dfrac{4}{5}$

6. 12% F $\dfrac{1}{3}$ G $\dfrac{1}{4}$ H $\dfrac{1}{5}$ J $\dfrac{1}{10}$

7. 19% A $\dfrac{1}{2}$ B $\dfrac{1}{3}$ C $\dfrac{1}{4}$ D $\dfrac{1}{5}$

Use a fraction to estimate the percent of each number.

8. 24% of 80

9. 73% of 44

10. 51% of 120

11. 32% of 90

____

12. 39% of 50

13. 12% of 40

14. 19% of 30

15. 58% of 200

____

Use 1% or 10% to estimate the percent of each number.

16. 39% of 25

17. 11% of 20

18. 3% of 200

19. 71% of 50

____

20. 9% of 40

21. 49% of 60

22. 4% of 50

23. 89% of 30

____

Holt McDougal Mathematics

LESSON 6-3 | Practice B
Estimating with Percents

Use a fraction to estimate the percent of each number.

1. 21% of 82

2. 35% of 42

3. 47% of 164

4. 9% of 68

5. 65% of 78

6. 11% of 92

7. 26% of 124

8. 89% of 51

9. 77% of 198

10. 5% of 75

11. 31% of 148

12. 53% of 539

13. In 2004, about $38 out of every $100 spent on advertising was spent on television advertising. The amount spent on radio advertising was about 21% as much as was spent on television advertising. How much of every $100 spent on advertising was spent on radio advertising? ____________

Use 1% or 10% to estimate the percent of each number.

14. 32% of 46

15. 81% of 36

16. 15% of 44

17. 21% of 62

18. 3% of 72

19. 62% of 88

20. 12% of 48

21. 65% of 124

22. 18% of 147

23. 5% of 837

24. 37% of 213

25. 2% of 188

26. The Fresh Acres Swim Club has a $35,000 budget for pool maintenance this year. The club members have agreed to raise the budget by 4%. Estimate the pool maintenance budget for next year. ____________

 Holt McDougal Mathematics

LESSON 6-3 Practice C
Estimating with Percents

Use a fraction to estimate the percent of each number.

1. 23% of 53   2. 35% of 37   3. 46% of 113   4. 8% of 39

_______   _______   _______   _______

5. 43% of 57   6. 13% of 127   7. 61% of 535   8. 53% of 175

_______   _______   _______   _______

Use 1% or 10% to estimate the percent of each number.

9. 37% of 63   10. 93% of 57   11. 16% of 83   12. 22% of 45

_______   _______   _______   _______

13. 5% of 91   14. 13% of 65   15. 58% of 135   16. 41% of 543

_______   _______   _______   _______

Every Tuesday, the Build-It-Yourself Warehouse gives senior citizens a 5% discount. Use estimation to complete the chart below to find the approximate cost of these items purchased on a Tuesday by a senior citizen.

	Item	Original Price	Estimated Discount	Estimated Final Price
17.	Electric drill	$45.99		
18.	Screwdriver set	$17.99		
19.	Rake	$23.50		
20.	Flower box	$37.79		
21.	Power saw	$95.29		
22.	10 lb bag of grass seed	$12.75		

    Holt McDougal Mathematics

LESSON 6-3 — Review for Mastery
Estimating with Percents

To estimate the percent of a number, choose a fraction that is close to the given percent.

Percent	5%	10%	20%	25%	$33\frac{1}{3}$%	40%	50%	60%	75%	80%
Fraction	$\frac{1}{20}$	$\frac{1}{10}$	$\frac{1}{5}$	$\frac{1}{4}$	$\frac{1}{3}$	$\frac{2}{5}$	$\frac{1}{2}$	$\frac{3}{5}$	$\frac{3}{4}$	$\frac{4}{5}$

Estimate 27% of 123.

27% is about 25%, which is equivalent to $\frac{1}{4}$.

123 rounded to the nearest ten is 120.

$\frac{1}{4} \cdot 120 = 30$

So, 27% of 123 is about 30.

Use a fraction to estimate the percent of each number.

1. 17% of 49

   17% is about ____, which is ____.

   49 rounded to the nearest ten is ____.

   ____ • ____ = ____

   So, 17% of 49 is about ____.

2. 58% of 298

   58% is about ____, which is ____.

   298 rounded to the nearest ten is ____.

   ____ • ____ = ____

   So, 58% of 298 is about ____.

3. 4% of 42

   4% is about ____, which is ____.

   42 rounded to the nearest ten is ____.

   ____ • ____ = ____

   So, 4% of 42 is about ____.

4. 46% of 533

   46% is about ____, which is ____.

   533 rounded to the nearest ten is ____.

   ____ • ____ = ____

   So, 46% of 533 is about ____.

5. 11% of 88

   ____

6. 74% of 203

   ____

7. 21% of 346

   ____

8. 79% of 55

   ____

Holt McDougal Mathematics

Review for Mastery

LESSON 6-3

Estimating with Percents (continued)

You can also choose a percent that is close to the given percent and use combinations of smaller percents.

1% of a number	Multiply by 0.01.
5% of a number	Find 10% of the number and divide by 2.
10% of a number	Multiply by 0.1.

Estimate 68% of 383.

68% is about 70%.

70% = 7 • 10%

383 rounded to the nearest ten is 380.

10% of 380 is 38, which is about 40.

7 • 40 = 280

So, 70% of 383 is about 280.

Use 1% or 10% to estimate the percent of each number.

9. 15% of 278

   15% = _____ + _____

   278 rounded to the nearest ten _____

   _____% of 280 = _____.

   _____% of 280 = _____.

   _____ + _____ = _____

   So, 15% of 278 is about _____.

10. 61% of 124

    61% is about _____%.

    _____% = _____ • _____ %

    124 rounded to the nearest ten _____.

    _____% of _____ = _____.

    _____ • _____ = _____

    So, 61% of 124 is about _____.

11. 18% of 62

____

12. 42% of 179

____

13. 5% of 317

____

14. 79% of 145

____

Holt McDougal Mathematics

<table><tr><td>LESSON
6-3</td><td></td></tr></table>

Challenge
The Octagon Garden

Estimate each percent. Find your way through the octagon garden by moving to the neighboring octagon with the next greater value. Keep going until you find your way out.

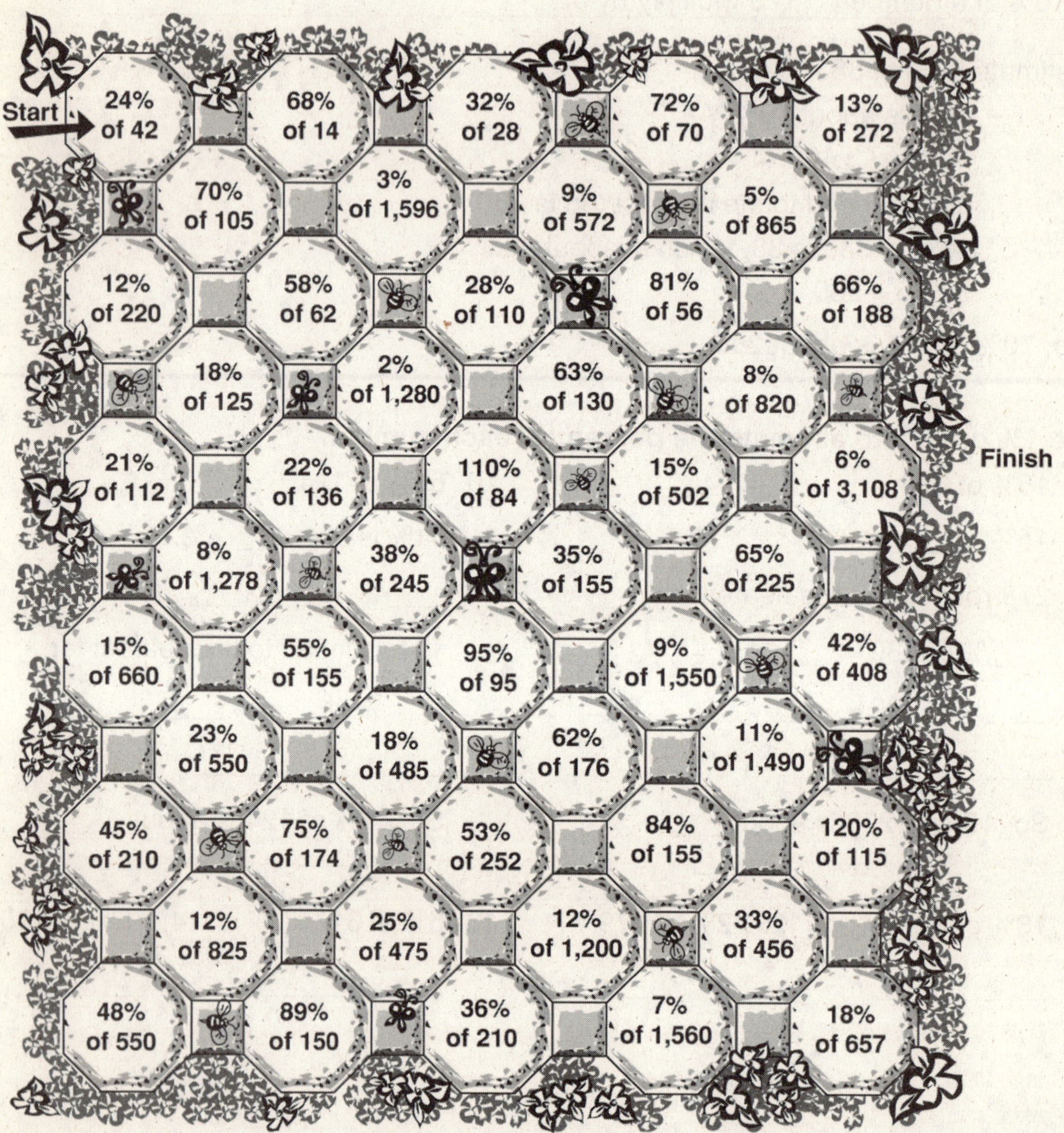

 Holt McDougal Mathematics

Problem Solving

LESSON 6-3

Estimating with Percents

Write the correct answer.

Use the graph to solve Exercises 1–3.

1. If a ski resort in the Rocky Mountain region had 125 visitors, about how many would be snowboarders?

____

2. Recently at one ski resort, 120 out of 400 guests were snowboarders. In which region is this resort?

____

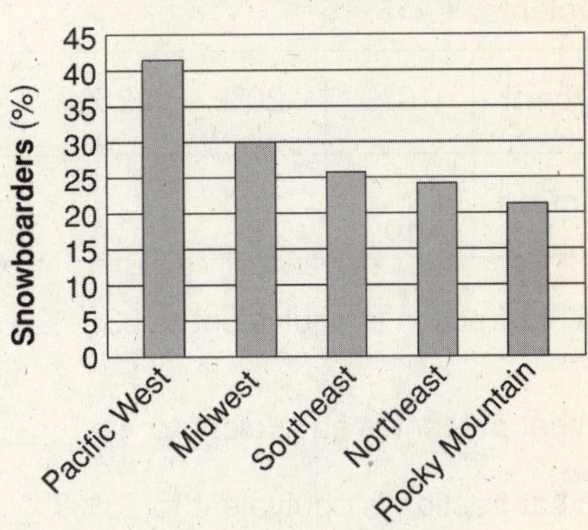

3. Last year, a ski resort in the Northeast region had an average of 556 visitors per weekend. About how many of them were snowboarders?

____

4. A large department store has an average of 3,456 shoppers per day. About 26% of the shoppers buys something in the store. About how many shoppers each day spend money in the store?

____

Choose the letter for the best answer.

5. LaToya bought a new car for $29,000. She is entitled to an 11% rebate. About how much will the car cost after the rebate?

   A $3,000   C $22,000

   B $10,000   D $26,000

6. The cost of a video game is $29.95. Sales tax is 6%. About how much will the video game cost, including tax?

   F about $29   H about $32

   G about $30   J about $35

7. In 2004, the Brooklyn Public Library spent about $7.6 million on acquisitions. The Detroit Public Library spent about 26% of that amount. About how much money did the Detroit Public Library spend?

   A $2 million   C $8 million

   B $3 million   D $26 million

8. In 2002, the average starting salary of a high school teacher in Switzerland was $48,704. The average starting salary of a high school teacher in the United States was about 61% of that amount. About how much was the average starting salary in the U.S?

   F $6,000   H $30,000

   G $24,000   J $60,000

Holt McDougal Mathematics

LESSON 6-3 — Reading Strategies
Using a Table

The table shows commonly used percents and their fraction equivalents.

Percent	10%	20%	25%	$33\frac{1}{3}\%$	50%
Fraction	$\frac{1}{10}$	$\frac{1}{5}$	$\frac{1}{4}$	$\frac{1}{3}$	$\frac{1}{2}$

1. What fraction is equivalent to $33\frac{1}{3}\%$? _______________

2. What percent is equivalent to $\frac{1}{10}$? _______________

3. What fraction is equivalent to 25%? _______________

Estimation is used when an exact answer is not needed.
Estimating with percents is a useful skill.

**Write *exact answer* or *estimate* to show which method
is best for the situation.**

4. You are trying to decide if you have enough money to buy a
   new CD.

   ____

5. You are asked how much you would charge to mow a
   neighbor's yard.

   ____

6. You are figuring the discount on a sale item.

   ____

**The table of percent and fraction equivalents can help you
make good estimates. Answer each question.**

7. What percent is close to 30%? _______________

8. What fraction is close to 45%? _______________

9. What percent is close to 8%? _______________

Holt McDougal Mathematics

LESSON 6-3 — Puzzles, Twisters & Teasers

Learn All About It!

Choose the better estimate. Then answer the riddle.

S   52% of 234   about   117 or 85.

T   31% of 180   about   36 or 54.

A   18% of 150   about   30 or 50.

E   8% of 310   about   31 or 16.

C   11% of 99   about   15 or 10.

H   23% of 98   about   20 or 30

Y   8% of 261   about   26 or 16.

W   41% of 16   about   6 or 10.

P   5% of 82   about   8 or 4.

L   52% of 83   about   31 or 41.

O   10% of 48   about   5 or 10.

R   39% of 19   about   4 or 8.

Why did all the cats jump off the fence?

___ ___ ___ ___   ___ ___ ___ ___   ___ ___ ___
54   20   31   26   6   31   8   31   30   41   41

___ ___ ___ ___ ___ ___ ___ ___
10   5   4   26   10   30   54   117

LESSON 6-4 Practice A
Percent of a Number

45	72	12	27	19	1.2	129	1.9	180
343	128	24	0.19	216	60	81	6	75

Find the percent of each number. Match each answer with a number in the box.

1. 60% of 40

2. 15% of 80

3. 5% of 120

4. 1% of 190

5. 120% of 50

6. 43% of 300

7. 150% of 30

8. 36% of 600

9. 9% of 800

10. 200% of 64

11. 27% of 300

12. 225% of 12

Find the percent of each number. Check whether your answer is reasonable.

13. 20% of 75

14. 25% of 64

15. 4% of 75

16. 125% of 60

17. 2% of 400

18. 160% of 80

19. 12% of 50

20. 230% of 70

21. 87% of 500

22. 28% of 250

23. 500% of 25

24. 3% of 300

25. The highest recorded temperature in Alaska occurred in 1915 at 100 °F. The highest recorded temperature in Indiana occurred in 1936 and was 16% higher than the highest temperature in Alaska. What is the highest recorded temperature in Indiana?

____

Holt McDougal Mathematics

Practice B

LESSON 6-4

Percent of a Number

Find the percent of each number.

1. 25% of 56 2. 10% of 110 3. 5% of 150 4. 90% of 180

_______________ _______________ _______________ _______________

5. 125% of 48 6. 225% of 88 7. 2% of 350 8. 285% of 200

_______________ _______________ _______________ _______________

9. 150% of 125 10. 46% of 235 11. 78% of 410 12. 0.5% of 64

_______________ _______________ _______________ _______________

Find the percent of each number. Check whether your answer is reasonable.

13. 55% of 900 14. 140% of 50 15. 75% of 128 16. 3% of 600

_______________ _______________ _______________ _______________

17. 16% of 85 18. 22% of 105 19. 0.7% of 110 20. 95% of 500

_______________ _______________ _______________ _______________

21. 3% of 750 22. 162% of 250 23. 18% of 90 24. 23.2% of 125

_______________ _______________ _______________ _______________

25. 0.1% of 950 26. 11% of 300 27. 52% of 410 28. 250% of 12

_______________ _______________ _______________ _______________

29. The largest frog in the world is the goliath, found in West Africa.
This type of frog can grow to be 12 inches long. The smallest
frog in the world is about 4% as long as the goliath. What is the
approximate length of the smallest frog in the world?

____

  Holt McDougal Mathematics

LESSON 6-4 — Practice C
Percent of a Number

Find the percent of each number. Round answers to the nearest tenth, if necessary.

1. 67% of 131

2. 19% of 73

3. 7% of 129

4. 1.1% of 67

5. 128% of 75

6. 227% of 183

7. 159% of 271

8. 271% of 159

9. 17% of 91

10. 23% of 77

11. 7% of 153

12. 7.3% of 73

13. 1.95% of 65

14. 137% of 81

15. 0.03% of 532

16. 541% of 132

Solve.

17. What number is 55% of 250?

18. 39% of 115 is what number?

19. 6% of 95 is what number?

20. What number is 80% of 860?

21. What number is 41% of 308?

22. 125% of 425 is what number?

23. 300% of 57 is what number?

24. What number is 180% of 110?

25. In 2004, there were 10,649 commercial radio stations in the United States. About 19.22% of those stations were devoted to country music. About how many country music radio stations were there in 2004?

Holt McDougal Mathematics

LESSON 6-4

Review for Mastery

Percent of a Number

You can use this proportion to solve percent problems.

$$\frac{\text{part}}{\text{total}} = \frac{\text{percent}}{100}$$

Find 20% of 65.

$$\frac{x}{65} = \frac{20}{100}$$

$100 \cdot x = 65 \cdot 20$ Find the

$100x = 1{,}300$ cross products.

$$\frac{100x}{100} = \frac{1{,}300}{100}$$ Simplify.

$$x = 13$$

So, 20% of 65 is 13.

> 20 is the **percent,** and 65 is the **total.** So, the **part** is unknown.

Find the percent of each number.

1. 40% of 90

   a. part = _________

   b. total = _________

   c. percent = _________

   d. $\dfrac{x}{\rule{1cm}{0.4pt}} = \dfrac{\rule{1cm}{0.4pt}}{100}$

   e. 100x = _________

   f. x = _________

2. 85% of 520

   a. part = _________

   b. total = _________

   c. percent = _________

   d. $\dfrac{x}{\rule{1cm}{0.4pt}} = \dfrac{\rule{1cm}{0.4pt}}{\rule{1cm}{0.4pt}}$

   e. 100x = _________

   f. x = _________

3. 30% of 80

4. 47% of 300

5. 45% of 200

6. 120% of 70

7. 65% of 40

8. 25% of 76

9. 115% of 40

10. 275% of 12

Holt McDougal Mathematics

LESSON 6-4

Challenge

Percent Puzzler

Use the percent clues to find each number below.

1. This number is 8 more than $83\frac{1}{2}$% of 20. __________

2. This number is 5 less than 28% of 292. __________

3. This number is equal to the sum of 4% of 75 and $32\frac{1}{2}$% of 64. __________

4. This number is 7.82 more than $15\frac{3}{4}$% of 320. __________

5. This number is equal to the sum of 52% of 86 and 35% of 52. __________

6. This number is equal to the product of 8% of 75 and 12% of 60. __________

7. This number is equal to the sum of 5% of 125 and $55\frac{1}{2}$% of 20. __________

8. This number is equal to the product of 24% of 225 and $3\frac{1}{5}$% of 45. __________

9. This number is equal to the sum of $9\frac{3}{10}$% of 110 and $38\frac{1}{4}$% of 124. __________

10. This number is 23.04 less than $14\frac{1}{2}$% of 768. __________

11. This number is 50.21 more than the sum of
    $15\frac{1}{2}$% of 42 and 64% of 112. __________

12. This number is equal to the difference of
    150% of 85 and 280% of 0.3. __________

13. This number is equal to the square of $7\frac{1}{2}$% of 160. __________

14. This number is 6.8 less than the product of 5 and 8% of 517. __________

Holt McDougal Mathematics

LESSON 6-4 Problem Solving
Percent of a Number

Write the correct answer.

The world population is estimated to exceed 9 billion by the year 2050. Use the circle graph to solve Exercises 1–3.

1. What is the estimated population of Africa in the year 2050?

   ____

2. Which continent is estimated to have more than 5.31 billion people by the year 2050?

   ____

3. In the year 2002, the world population was estimated at 6 billion people. Based on research from the World Bank, about 20% lived on less than $1 per day. How many people lived on less than $1 per day?

   ____

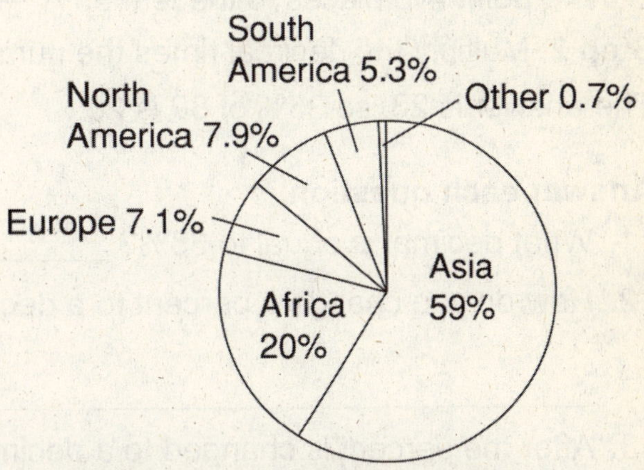

4. What is the combined estimated population for North and South America by the year 2050?

   ____

Choose the letter for the best answer.

5. The student population at King Middle School is 52% female. The total student population is 1,225 students. How many boys go to King Middle School?

   A 538 boys   C 637 boys

   B 588 boys   D 1,173 boys

6. There are 245 students in the seventh grade. If 40% of them ride the bus to school, how many seventh graders do not ride the bus to school?

   F 98 students   H 185 students

   G 147 students   J 205 students

7. A half-cup of pancake mix has 5% of the total daily allowance of cholesterol. The total daily allowance of cholesterol is 300 mg. How much cholesterol does a half-cup of pancake mix have?

   A 100 mg   C 20 mg

   B 60 mg   D 15 mg

8. Carey needs $45 to buy her mother a birthday present. She has saved 22% of the amount so far. How much more does she need?

   F $35.10   H $23.00

   G $44.01   J $9.90

LESSON 6-4

Reading Strategies
Find a Pattern

You can use decimals to find the percent of a number.

Find 35% of 80.

35% of 80 means 35% times 80.

Step 1: Change the percent to a decimal by moving the decimal point two places to the left. 35% → 0.35

Step 2: Multiply the decimal times the number. → 0.35 × 80

The answer is 28, so 35% of 80 is 28.

Answer each question.

1. What decimal is equal to 35%? _________________

2. How do you change a percent to a decimal?

   ____

3. After the percent is changed to a decimal, what is the next step in finding the percent of a number?

   ____

4. Write 10% as a decimal. _________________

5. What is 10% of 60? _________________

6. What is 20% of 60? _________________

7. What is 30% of 60? _________________

8. What is 40% of 60? _________________

9. What pattern did you notice in the answers to 10, 20, 30, and 40 percent of 60?

   ____

Holt McDougal Mathematics

LESSON 6-4	**Puzzles, Twisters & Teasers**

Hi and Dry!

Decide whether or not each equation is correct. Circle the letters above your answers. Then solve the riddle.

1. 7% of 200 = 21

   B   **Y**

  correct   incorrect

2. 16% of 50 = 18

   FO

  correct   incorrect

3. 8% of 50 = 4

   U   **G**

  correct   incorrect

4. 5% of 12 = 0.06

   K   **R**

  correct   incorrect

5. 67% of 90 = 60.3

   S   **L**

  correct   incorrect

6. 16% of 70 = 1.17

   M   **H**

  correct   incorrect

7. 20% of 65 = 15

   N   **A**

  correct   incorrect

8. 0.5% of 36 = 0.18

   D   **P**

  correct   incorrect

9. 145% of 210 = 304.5

   O   **T**

  correct   incorrect

10. 1% of 4 = 4

   ZW

  correct   incorrect

What can fall on the sea without getting wet?

___ ___ ___ ___ ___

____

LESSON 6-5   Practice A
Solving Percent Problems

Solve. Choose the letter for the best answer.

1. 50 is 50% of what number?

    A 25   C 75

    B 50   D 100

2. 25% of what number is 10?

    F 2.5   H 40

    G 15   J 250

3. What percent of 50 is 10?

    A 5%   C 20%

    B 10%   D 60%

4. 9 is what percent of 12?

    F 125%   H 33%

    G 75%   J 30%

5. 25 is 20% of what number?

    A 5   C 95

    B 45   D 125

6. 75% of what number is 15?

    F 20   H 60

    G 25   J 90

7. 24 is what percent of 40?

    A 10%   C 60%

    B 16%   D 64%

8. What percent of 36 is 9?

    F 3%   H 45%

    G 25%   J 75%

9. What percent of 80 is 48?

____

10. What percent of 50 is 5?

____

11. What percent of 20 is 16?

____

12. What percent of 25 is 15?

____

13. 11 is 50% of what number?

____

14. 40 is 20% of what number?

____

15. 30 is 25% of what number?

____

16. 150 is 200% of what number?

____

17. 60% of what number is 84?

____

18. 125 is 25% of what number?

____

19. The sales tax on a $150 DVD player is $6. What is the sales tax rate?

____

       Holt McDougal Mathematics

Practice B
LESSON 6-5

Solving Percent Problems

Solve.

1. 50 is 40% of what number?

2. 12 is 25% of what number?

3. 18 is what percent of 60?

4. 12 is what percent of 96?

5. 4% of what number is 25?

6. 80% of what number is 160?

7. What percent of 55 is 22?

8. What percent of 75 is 6?

9. 15 is 30% of what number?

10. 8% of what number is 2?

11. 7 is what percent of 105?

12. 24 is 40% of what number?

13. 10% of what number is 14?

14. 16 is what percent of 200?

15. What percent of 32 is 4?

16. What percent of 150 is 60?

17. 1% of what number is 11?

18. 20% of what number is 14?

19. The sales tax on a $750 computer at J & M Computers is $48.75. What is the sales tax rate?

20. A hardcover book sells for $24 at The Bookmart. Ben pays a total of $25.02 for the book. What is the sales tax rate?

     Holt McDougal Mathematics

LESSON 6-5 Practice C
Solving Percent Problems

Solve. Round answers to the nearest tenth, if necessary.

1. 14 is 28% of what number?

2. 28% of what number is 7?

3. 35% of what number is 70?

4. What percent of 63 is 7?

5. 28 is what percent of 210?

6. 65% of what number is 13?

7. What percent of 180 is 36?

8. 15 is 45% of what number?

9. 7 is 2% of what number?

10. What percent of 144 is 108?

11. What percent of 350 is 840?

12. 13 is what percent of 77?

13. 11 is 4% of what number?

14. 135% of what number is 115?

15. 14 is 16% of what number?

16. 12 is 30% of what number?

17. 18 is what percent of 108?

18. 9 is what percent of 39?

19. I am thinking of a number. This number is 18% of 425. What is the number?

20. I am thinking of a percent. This percent times 135 gives 40.5 as the result. What is the percent?

21. Jed buys a sweater that has a sale price of $48. Jed gives the sales clerk $60 and receives $7.80 in change. What is the sales tax rate?

Holt McDougal Mathematics

LESSON 6-5

Review for Mastery

Solving Percent Problems

You can use this proportion to solve percent problems.

$$\frac{part}{total} = \frac{percent}{100}$$

9 is what percent of 12?

$$\frac{9}{12} = \frac{x}{100}$$

$$12 \cdot x = 9 \cdot 100$$

$$12x = 900$$

$$\frac{12x}{12} = \frac{900}{12}$$

$$x = 75$$

So, 9 is 75% of 12.

> 9 is the **part**, and 12 is the **total**. So, the **percent** is unknown.

24 is 30% of what number?

$$\frac{24}{x} = \frac{30}{100}$$

$$30 \cdot x = 24 \cdot 100$$

$$30x = 2{,}400$$

$$\frac{30x}{30} = \frac{2{,}400}{30}$$

$$x = 80$$

So, 24 is 30% of 80.

> 24 is the **part**, and 30 is the **percent**. So, the **total** is unknown.

Solve.

1. What percent of 25 is 14?

   a. part = _______

   b. total = _______

   c. percent = _______

   d. $\dfrac{}{} = \dfrac{x}{100}$

   e. _______ $x =$ _______

   f. $x =$ _______

   g. 14 is _______ of 25.

2. 80% of what number is 16?

   a. part = _______

   b. total = _______

   c. percent = _______

   d. $\dfrac{}{x} = \dfrac{}{100}$

   e. _______ $x =$ _______

   f. $x =$ _______

   g. 16 is 80% of _______.

3. What percent of 20 is 11?

   ____

4. 18 is 45% of what number?

   ____

5. 15 is what percent of 5?

   ____

6. 75% of what number is 105?

   ____

Holt McDougal Mathematics

LESSON 6-5 Challenge
Percentile Rank

Just as there are three quartiles (the lower quartile, the median, and the upper quartile) that divide a data set into four equal groups, there are 99 *percentiles* that divide a data set into 100 groups.

The definition of a percentile is:

$$\text{percentile of score } x = \frac{\text{number of scores less than or equal to score}}{\text{total number of scores}} \cdot 100$$

The frequency table at the right shows the test scores for 28 students.

Find the percentile corresponding to 80.

$$\text{percentile of } 80 = \frac{\text{number of scores less than or equal to 80}}{\text{total number of scores}} \cdot 100$$

$$= \frac{14}{28} \times 100 = 0.5 \times 100 = 50$$

So, 80 is the 50th percentile.

Score	Frequency
100	1
95	2
90	5
85	6
80	7
75	3
70	2
65	2

Use the frequency table to find the percentile corresponding to each score. Round your answer to the nearest whole number.

1. 90

2. 70

3. 100

4. 75

5. 95

6. 85

Use the test scores listed below to find the percentile corresponding to each score. Round your answer to the nearest whole number. (*Hint:* Make a frequency table of the scores.)

84, 77, 77, 77, 92, 77, 84, 84, 95, 84, 68, 92, 84, 100, 77,
77, 84, 92, 77, 92, 92, 95, 77, 68, 84, 100, 92, 84, 95, 92

7. 100

8. 95

9. 92

10. 84

11. 77

12. 68

Holt McDougal Mathematics

LESSON 6-5 Problem Solving
Solving Percent Problems

Write the correct answer.

1. At one time during 2001, for every 20 copies of *Harry Potter and the Sorcerer's Stone* that were sold, 13.2 copies of *Harry Potter and the Prisoner of Azkaban* were sold. Express the ratio of copies of *The Prisoner of Azkaban* sold to copies of *The Sorcerer's Stone* sold as a percent.

____

2. A souvenir mug sells for $8.00 at a hotel gift store. Kendra gives the clerk $9.00 and receives $0.56 in change. What is the sales tax rate?

____

3. Craig just finished reading 120 pages of his history assignment. If the assignment is 125 pages, what percent has Craig read so far?

____

4. Hal's Sporting Goods had a 1-day sale. The original price of a mountain bike was $325. On sale, it was $276.25. What is the percent reduction for this sale?

____

Choose the correct letter for the best answer.

5. China's area is about 3.7 million square miles. It is on the continent of Asia, which has an area of about 17.2 million square miles. About what percent of the Asian continent does China cover?

   A about 10%   C about 17%

   B about 15%   D about 22%

6. After six weeks, the tomato plant that was given extra plant food and water was 26 centimeters tall. The tomato plant that was not given any extra plant food was only 74.5% as tall. How tall was the tomato plant that was not given extra plant food?

   F 1.94 cm   H 34.89 cm

   G 19.37 cm   J 48.5 cm

7. In a survey, 46 people, which was 20% of those surveyed, chose red as their favorite color. How many people were surveyed?

   A 66 people   C 460 people

   B 230 people   D 920 people

8. Of the 77 billion food and drink cans, bottles, and jars Americans throw away each year, about 65% of them are cans. How many food and drink cans, to the nearest billion, do Americans throw away each year?

   F 12 billion   H 50 billion

   G 17 billion   J 65 billion

    Holt McDougal Mathematics

LESSON 6-5

Reading Strategies
Connecting Words and Symbols

You can connect words and symbols to write equations for percent problems.

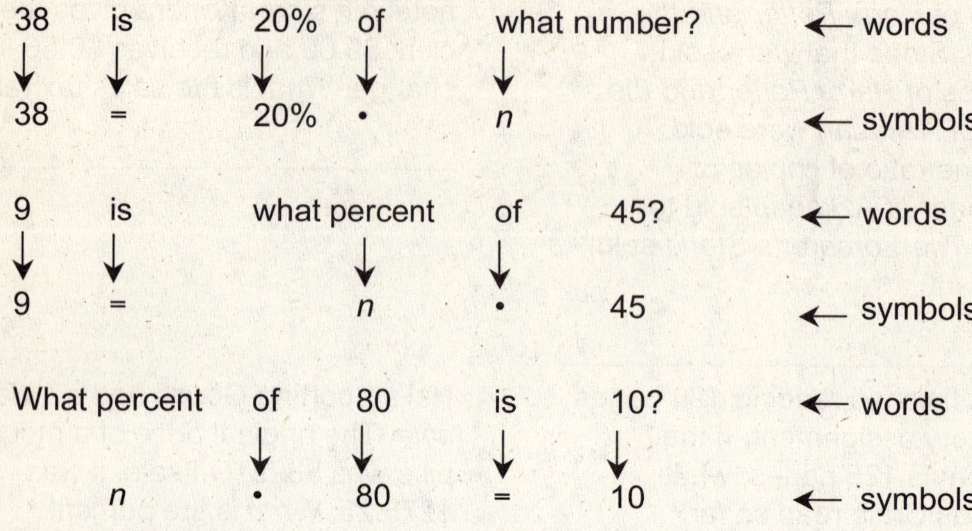

Answer each question.

1. What is the symbol for the word "of"?

   ____

2. What symbol means "is"?

   ____

3. What symbol in the above examples stands for the unknown number or percent?

   ____

4. Write the equation for this percent problem using symbols: 6 is 10% of what number?

   ____

5. Write the words for this equation: $27 = n \cdot 30$.

   ____

6. Write the symbols for this percent problem: What percent of 40 is 30?

   ____

   ____

Holt McDougal Mathematics

LESSON 6-5 — Puzzles, Twisters & Teasers
Rock On!

Fill in the blanks to complete each statement. Then solve the riddle.

L   25 is _______% of 100

R   27 is _______% of 30

A   105 is _______% of 75

O   16 is _______% of 48

V   56 is 140% of _______

C   40 is _______% of 60

I   10 is _______% of 80

K   9 is _______% of 45

T   6 is 10% of _______

F   7 is 14% of _______

S   30 is 15% of _______

E   18 is _______% of 6

What kind of concert do pebbles enjoy?

____   ____ ____ ____ ____
140   90   33.3̄ 66.6̄   20

____ ____ ____ ____ ____ ____ ____ ____
50   300   200   60   12.5   40   140   25

Family Letter

6B Applying Percents

Dear Family,

The student is learning to find percent of change. The **percent of change** is the amount, stated as a percent, that a number increases or decreases.

$$\text{percent of change} = \frac{\text{amount of change}}{\text{original amount}}$$

Find the percent of change if 24 is decreased to 20.

$24 - 20 = 4$	Find the amount of change.
percent of change $= \dfrac{4}{24}$	Substitute values into formula.
$\approx 0.1\overline{6}$	Divide.
$\approx 16.7\%$	Write the decimal as a percent.

The percent of decrease is about 16.7%.

Find the percent of change if 75 is increased to 120.

$120 - 75 = 45$	Find the amount of change.
percent of change $= \dfrac{45}{75}$	Substitute values into formula.
$= 0.60$	Divide.
$= 60\%$	Write the decimal as a percent.

The percent of increase is 60%.

The student will also be introduced to some real-world percent applications, such as finding a sale price or the amount of mark-up or discount.

Jack's Snacks buys pretzels from a food supplier for $0.25 per package and sells each at an 80% increase in price. What is the mark-up on each package?

Write an equation.

$80\% \cdot 0.25 = n$	
$0.80 \cdot 0.25 = n$	Write the percent as a decimal.
$0.20 = n$	Multiply.

The amount of mark-up on each package is $0.20.

Vocabulary

These are the math words we are learning:

interest the amount that is collected or paid for the use of money

percent of change the amount a number increases or decreases

percent of decrease the amount of change a number goes down

percent of increase the amount of change a number goes up

principal the amount of money deposited or borrowed

simple interest money paid only on the principal

Holt McDougal Mathematics

Family Letter

6B Applying Percents continued

Tires-4-All is having a 35% off sale on all tires. The average tire regularly sells for $51.99. Find the amount of the discount and the sale price of one tire.

Find the amount of discount.

35% • 51.99 = n	Write an equation.
0.35 • 51.99 = n	Write the percent as a decimal.
18.1965 = n	Multiply.
$18.20 ≈ n	Round to the nearest cent.

The discount is about $18.20.

Find the sale price.

$51.99 – $18.20 = $33.79	Subtract the discount from the regular price.

The sale price is about $33.79.

The student will solve problems involving simple interest. To do this, he or she will learn the following formula:

$$I = p \cdot r \cdot t$$

Interest = principal • rate • time

If p = $450, r = 8.5%, and t = 3 years, find the simple interest.

$I = p \cdot r \cdot t$	Use the simple interest formula.
$I = 450 \cdot 0.085 \cdot 3$	Substitute. Use 0.085 for 8.5%.
$I = 114.75$	Multiply.

The simple interest is $114.75.

Encourage the student to find real-world applications of percent concepts.

Sincerely,

Holt McDougal Mathematics

CHAPTER 6

At-Home Practice

6B Applying Percents

Find each percent of change. Round answers to the nearest tenth of a percent, if necessary.

1. 60 is decreased to 50

2. 35 is increased to 70

3. 18 is increased to 99

4. 85 is decreased to 25

5. Allison buys baskets for $2.25 each. She increases the price of the baskets 60% and then resells them at her flea-market booth. What is the mark-up on the baskets?

6. Tim's hardware store has a sale on hammers for 25% off the marked price. If the hammers are priced at $12.50, find the amount of the discount and the sale price.

Find each missing value.

7. $I = ?$, $p = \$800$, $r = 8\%$, $t = 3$ years

8. $I = ?$, $p = \$16,000$, $r = 3\%$ $t = 7$ years

9. $I = \$720$, $p = \$800$, $r = 9\%$, $t = ?$ years

10. $I = \$157.50$, $p = ?$, $r = 5\%$ $t = 7$ years

11. Aaron invests $3,600 for 5 years at a simple interest rate of 6%. How much money does he receive in interest?

12. Shirley receives $5,000 per year from the simple interest on her money market funds. She has an annual interest rate of 4%. How much money does she have invested in her funds?

Answers: 1. 16.7% **2.** 100% **3.** 450% **4.** 70.6% **5.** $1.35 **6.** $3.13 discount; $9.37 sale price **7.** $192 **8.** $3,360 **9.** 10 years **10.** $450 **11.** $1,080 **12.** $125,000

 Holt McDougal Mathematics

Family Fun

Be a Smart Shopper

Materials

10 discount cards
10 product cards

Directions

- Cut out the cards. Each team randomly chooses 3 discount cards and 5 product cards.

- Each team must use each of their discount cards at least once on a product.

- Each team applies a discount to their products and totals up the discount amount.

- The team that saves the most money wins that round.

- The winner is the team who wins the best out of 3 rounds.

5%	10%	15%	20%	25%
30%	50%	5%	10%	40%
Music CD $16.50	Electronic Game System $249.99	Gas Card $25	Tropical Fish $30	Scooter $65.99
Mountain Bike $269	Baseball Glove $55.79	DVD Set $29.99	Notebook Paper $3.49	4-Piece Table Setting $89.99

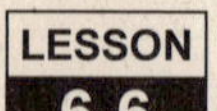

Practice A

Percent of Change

Complete the table.

	Problem	Amount of Change	Original Amount	% of Change
1.	25 is decreased to 17			
2.	24 is increased to 36			
3.	50 is decreased to 40			
4.	40 is increased to 56			

Find each percent of change. Round answers to the nearest tenth, if necessary.

5. 60 is decreased to 15 ______________

6. 15 is increased to 21 ______________

7. 12 is increased to 48 ______________

8. 100 is decreased to 25 ______________

9. 60 is decreased to 40 ______________

10. 80 is increased to 152 ______________

11. 15 is increased to 24 ______________

12. 72 is decreased to 64 ______________

13. The Big Bike Shop has in-line skates on sale for 15% off the regular price. A pair of in-line skates usually costs $89. Find the amount of the discount and the sale price.

____

14. In 1935, there were 15,295 banks in the United States. In 2003, there were 9,182 banks. What is the percent of change? Is it a percent increase or decrease?

____

15. A jewelry store is having a going-out-of-business sale. All merchandise is marked 40% off the regular price. The regular price of a watch is $59.95. Find the amount of the discount and the sale price.

____

16. Last year, there were 381 students at Woodland Middle School. This year, there are 419 students. What is the percent of change? Is it a percent increase or decrease?

____

Holt McDougal Mathematics

<table><tr><td>LESSON
6-6</td><td></td></tr></table>

Practice B
Percent of Change

Find each percent of change. Round answers to the nearest tenth, if necessary.

1. 20 is decreased to 11 _______________

2. 24 is increased to 30 _______________

3. 56 is decreased to 14 _______________

4. 25 is increased to 100 _______________

5. 18 is increased to 45 _______________

6. 90 is decreased to 75 _______________

7. 126 is decreased to 48 _______________

8. 65 is increased to 144 _______________

9. 42 is increased to 72 _______________

10. 84 is decreased to 8 _______________

11. 95 is increased to 145 _______________

12. 248 is decreased to 200 _______________

13. 105 is decreased to 32 _______________

14. 75 is increased to 350 _______________

15. 93 is decreased to 90 _______________

16. 16 is decreased to 2 _______________

17. A backpack that normally sells for $39 is on sale for 33% off. Find the amount of the discount and the sale price.

____

18. A sporting goods store is having a closeout on a certain style of running shoes. They are marked 55% off the regular price. The regular price is $79.95. Find the amount of the discount and the sale price.

____

19. A gallery owner purchased a very old painting for $3,000. The painting sells at a 325% increase in price. What is the retail price of the painting?

____

20. In August, the Simons' water bill was $48. In September, it was 15% lower. What was the Simons' water bill in September?

____

     Holt McDougal Mathematics

<table>
<tr><td>**LESSON**
6-6</td><td></td></tr>
</table>

Practice C
Percent of Change

Find each percent of change or amount of change. Round answers to the nearest tenth, if necessary.

1. 50 is decreased to 19 ______________

2. 25 is increased by 80% ______________

3. 40 is decreased to 33 ______________

4. 75 is increased to 200 ______________

5. $80 is increased by 125% ______________

6. $14.40 is decreased by 20% ______________

7. 11 is decreased by 10% ______________

8. 5.7 is increased to 7.8 ______________

9. 42.6 is increased to 71.2 ______________

10. 98 is increased by 85% ______________

11. 145 is decreased to 105 ______________

12. 79 is increased to 133 ______________

13. 40 is increased by 12.5% ______________

14. 119 is increased to 277 ______________

15. 537 is decreased to 289 ______________

16. 54 is decreased by 75% ______________

17. The school bought 6 new computers for the computer lab. Each computer has a retail price of $979. The school received a discount of 12% on each computer. Find the amount of the discount and the final price for all 6 computers.

____

18. Ali paid $85 for an antique clock several years ago. She recently found that it had increased in value 165%. What is the present value of the clock?

____

19. In 1950, there were 40.3 million cars registered in the United States. In 1990, there were 133.7 million cars registered in the United States. What was the percent increase of registered cars from 1950 to 1990?

____

Holt McDougal Mathematics

Name _________________________ Date _______________ Class _______________

Review for Mastery

Percent of Change

A change in a quantity is often described as a percent increase or
percent decrease. To calculate a percent increase or decrease, use
this equation.

$$\text{percent of change} = \frac{\text{amount of increase or decrease}}{\text{original amount}} \cdot 100$$

Find the percent of change from 28 to 42.
- First, find the amount of the change. $42 - 28 = 14$
- What is the original amount?   28
- Use the equation. $\dfrac{14}{28} \cdot 100 = 50\%$

An increase from 28 to 42 represents a 50% increase.

Find each percent of change.

1. 8 is increased to 22

   amount of change $22 - 8 =$ _________

   original amount _________

   _________ $\cdot\ 100 =$ _________ %

2. 90 is decreased to 81

   amount of change $90 - 81 =$ _________

   original amount _________

   _________ $\cdot\ 100 =$ _________ %

3. 125 is increased to 200

   amount of change $200 - 125 =$ _________

   original amount _________

   _________ $\cdot\ 100 =$ _________ %

4. 400 is decreased to 60

   amount of change $400 - 60 =$ _________

   original amount _________

   _________ $\cdot\ 100 =$ _________ %

5. 64 is decreased to 48

6. 140 is increased to 273

7. 30 is decreased to 6

8. 15 is increased to 21

9. 7 is increased to 21

10. 320 is decreased to 304

Holt McDougal Mathematics

LESSON 6-6

Challenge
Double Percent Change

Solve.

1. Enrollment in the school soccer league was 340 last year. This year, it dropped 15%. What is the enrollment this year? If enrollment increases 15% next year, what will the enrollment be? Round to the nearest whole number.

2. Enrollment in the Wilderness Club increased 8% for each of the last two years. Enrollment two years ago was 175. What was the enrollment for each of the following two years? Round to the nearest whole number.

3. The toy store buys stuffed animals from the manufacturer for $2.40 each. The store then sells them at a 96% increase in price. What is the retail price of each animal? If the customer brings in a coupon for 20% off the retail price, how much will the stuffed animal cost?

4. Enrollment in the Jacksonville Township Basketball League was 425 last year. This year, the number enrolled increased 12%. If enrollment drops 12% next year, how many people will be enrolled in the league? Round to the nearest whole number.

5. The manager of a card store buys cards from the manufacturer for $0.65 each. The store regularly sells the cards at a markup of 130%. For a sale, the manager marks down half of the cards 15%. What is the price of the cards that are on sale?

6. Jamal wants to buy a video game. Two stores in the neighborhood offer the game at $49.95. Jamal has a coupon for 18% off the regular price at Fun Electronics. The other store, Gamers, is having a sale with 8% off everything in the store. Jamal also has a coupon for 10% off the sale price at Gamers. At which store should he buy the game? Explain.

   Holt McDougal Mathematics

<table><tr><td>**LESSON**
6-6</td><td></td></tr></table>

Problem Solving
Percent of Change

Write the correct answer.

1. In 2002, U.S. consumers bought about 8.1 million new cars. In 2003, that number decreased by about 6%. To the nearest hundred thousand, or tenth of a million, how many new cars did U.S. consumers buy in 2003?

2. In Union County, Florida, the 1990 census listed the population at 10,252. The 2000 census listed the population as 13,442. What percent increase is this to the nearest tenth of a percent?

3. World production of motor vehicles increased from about 60 million in 2002 to 62 million in 2003. What was the percent increase to the nearest percent?

4. Arthur's dog, Shep, used to weigh 158 pounds. The vet put him on a diet and he lost 13% of his weight. To the nearest pound, how much does Shep weigh now?

5. The number of volunteers rose from 47 on Monday to 64 on Tuesday. What is the percent increase to the nearest tenth of a percent?

6. Coretta's bowling average decreased from 158 to 133. What is the percent decrease to the nearest tenth of a percent?

Choose the correct letter for the best answer.

7. Shandra scored 75 on her first math test. She scored 20% higher on her next math test. What did she score on the second test?

   A 80   C 90

   B 85   D 95

8. Jim's Gym had income of $20,350 last month. The total increased by $2,000 this month. What was the percent increase to the nearest percent?

   F 11%   H 7%

   G 10%   J 6%

9. Last year, the average number of absences in school was 8 students per day. This year, the absentee rate is down to 6 students per day. What is the percent decrease in student absences this year?

   A 75%   C 25%

   B 33%   D 66%

10. During the 2002–2003 ski season $171 million worth of snowboarding equipment was sold. Sales increased by about 15% during the 2003–2004 season. About how much were sales of snowboarding equipment in the 2003–2004 season?

   F $145 million   H $186 million

   G $156 million   J $197 million

 Holt McDougal Mathematics

<table>
<tr><td>LESSON
6-6</td><td></td></tr>
</table>

Reading Strategies
Analyze Information

Percent can be used to describe change. It is shown as a ratio.

$$\text{Percent of change} = \frac{\text{amount of change}}{\text{original amount}}$$

The following steps describe how the percent of change is figured on
a savings account that starts with $50.

This is the original amount in the account:	$50
This is the current amount in the account:	$30
This is the amount that the account decreased by:	$20

$\text{Percent of change} = \dfrac{\$20}{\$50}$ the amount of change over the original amount

Savings went down, so this ratio is → the **percent of decrease** in savings.

1. How much money was placed into the savings account when it
   opened?

   ____

2. Did the number of dollars in the account increase or decrease?

   ____

3. When you are figuring the percent of change, where is the
   original number placed in the fraction?

   ____

**Use this information for Exercises 4–6: A clothing salesman sold
25 shirts his first day on the job and 45 shirts the second day.**

4. What is the original number of shirts he sold?

   ____

5. How many more shirts did he sell the second day than the first
   day?

   ____

6. Write the fraction that shows the amount of change over the
   original amount.

   ____

　　　　　　　　　　　　　　　　　　　　　　　Holt McDougal Mathematics

Puzzles, Twisters & Teasers

LESSON 6-6

Be Patient!

Circle words from the list in the word search. Then find a word that answers the riddle. Circle it and write it on the line.

percent	change	increase	decrease	round
tenth	substitute	decimal	discount	

```
S P E R C E N T E N T H O
R U D E C I M A L U J O D
O P B D I S C O U N T E E
U A R S U N J K O P L R C
N T A S T U P L N E Q X R
D I L K J I N C R E A S E
Q E S D F G T Y H J K O A
P N U H Y T G U W E T R S
B T X O K W I L T C G Y E
N S M I C H A N G E Q W R
```

When do dentists get angry?

When they run out of ___ ___ ___ ___ ___ ___ ___ .

LESSON 6-7   Practice A
Simple Interest

Find each missing value.

1. $p = \$1{,}000$, $r = 5\%$, $t = 2$ years

   $I =$ ________ $\cdot$ ________ $\cdot$ ________

   $I =$ ________________

2. $p = \$600$, $r = 4\%$, $t = 3$ years

   $I =$ ________ $\cdot$ ________ $\cdot$ ________

   $I =$ ________________

3. $I = \$330$, $r = 3\%$, $t = 1$ year

   ________ $= p \cdot$ ________ $\cdot$ ________

   $p =$ ________________

4. $I = \$270$, $r = 5\%$, $t = 3$ years

   ________ $= p \cdot$ ________ $\cdot$ ________

   $p =$ ________________

5. $I = \$600$, $p = \$2{,}500$, $t = 4$ years

   ________ $=$ ________ $\cdot r \cdot$ ________

   $r =$ ________________

6. $I = \$108$, $p = \$900$, $t = 3$ years

   ________ $=$ ________ $\cdot r \cdot$ ________

   $r =$ ________________

7. $p = \$250$, $r = 6\%$, $t = 5$ years

   $I =$ ________________

8. $p = \$3{,}000$, $r = 7\%$, $t = 4$ years

   $I =$ ________________

9. $I = \$750$, $r = 4\%$, $t = 5$ years

   $p =$ ________________

10. $I = \$696$, $r = 3\%$, $t = 4$ years

    $p =$ ________________

11. $I = \$425$, $p = \$1{,}700$, $t = 5$ years

    $r =$ ________________

12. $I = \$1{,}680$, $p = \$12{,}000$, $t = 2$ years

    $r =$ ________________

13. You deposit $5,000 in an account that earns 5% simple interest. How long will it be before the total amount is $6,000? ________

14. After 6 years, an account that earns 4% simple interest has earned $480 in interest. How much was the initial deposit? ________

15. A deposit of $7,500 earns $3,900 over a period of 8 years. What is the simple interest rate? ________

16. You deposit $4,500 in an account that earns 6% simple interest. How much will be in your account after 5 years? ________

    Holt McDougal Mathematics

Practice B

Simple Interest

Find each missing value.

1. $p = \$1,500$, $r = 5\%$, $t = 3$ years

   $I = $ _______________

2. $p = \$6,000$, $r = 4\%$, $t = 2$ years

   $I = $ _______________

3. $I = \$30$, $r = 4\%$, $t = 2$ years

   $p = $ _______________

4. $I = \$180$, $r = 5\%$, $t = 3$ years

   $p = $ _______________

5. $I = \$20$, $p = \$250$, $t = 2$ years

   $r = $ _______________

6. $I = \$144$, $p = \$800$, $t = 3$ years

   $r = $ _______________

7. $p = \$525$, $r = 3\%$, $t = 1$ year

   $I = $ _______________

8. $p = \$3,200$, $r = 6\%$, $t = 4$ years

   $I = $ _______________

9. $I = \$450$, $r = 6\%$, $t = 4$ years

   $p = $ _______________

10. $I = \$1,440$, $r = 3\%$, $t = 5$ years

    $p = $ _______________

11. $I = \$1,275$, $p = \$5,100$, $t = 5$ years

    $r = $ _______________

12. $I = \$3,920$, $p = \$14,000$, $t = 4$ years

    $r = $ _______________

13. $p = \$1,300$, $r = 4.5\%$, $t = 6$ months

    $I = $ _______________

14. $I = \$47.25$, $r = 3.5\%$, $t = 1.5$ years

    $p = $ _______________

15. $I = \$891$, $p = \$2,700$, $t = 5.5$ years

    $r = $ _______________

16. $I = \$126$, $p = \$400$, $t = 9$ years

    $r = $ _______________

17. You deposit $2,500 in an account that earns 4% simple interest. How long will it be before the total amount is $3,000? _______________

18. You deposit $5,000 in account that earns 6.5% simple interest. How much will be in the account after 3 years? _______________

19. A deposit of $10,000 was made to an account the year you were born. After 12 years, the account is worth $16,600. What simple interest rate did the account earn? _______________

20. How long will it take for $6,500 to double at a simple interest rate of 7%? Round to the nearest tenth of a year. _______________

LESSON 6-7 Practice C

Simple Interest

Complete the table.

	Principal	Interest Rate	Time	Simple Interest
1.	$2,000	3.5%	3 years	
2.	$1,250		4 years	$350
3.	$500	4.5%		$157.50
4.		5%	54 months	$2,115
5.	$12,000		2 years	$1,080
6.	$1,800	7.5%	6 months	
7.		6%	4 years	$73.80
8.	$8,500	6.5%		$6,630
9.		3.5%	5 years	$700
10.	$3,300	4.75%		$313.50
11.	$6,800		16 years	$2,720
12.	$2,400	5.5%	30 months	

Solve.

13. A deposit of $500 in an account earns 6% simple interest. How long will it be before the total amount is $575? _______

14. What simple interest rate is needed for $1,000 to grow to $1,071.25 in 9 months? _______

15. A deposit of $3,000 becomes $3,810 after 6 years. What is the simple interest rate on the account? _______

16. How long will it take for $1,000 to double at a simple interest rate of 5.5%?

____

17. Consuelo deposited an amount of money in a savings account that earned 6.3% simple interest. After 20 years, she had earned $5,922 in interest. What was her initial deposit? _______

18. A deposit of $2,500 grew to $3,325 after 6 years. What is the final value of a deposit of $7,500 at the same interest rate for the same period of time? _______

  Holt McDougal Mathematics

Review for Mastery

LESSON 6-7

Simple Interest

When you put money into a bank account, you may receive simple interest for loaning the bank your money.

$$Interest = Principal \cdot Rate \cdot Time$$
$$I = p \cdot r \cdot t$$

You can use the expression $\dfrac{I}{p \cdot r \cdot t}$ to solve interest problems.

- To find interest (I), put your finger over I. Perform the operations for letters you see.

- To find principal (p), put your finger over p. Perform the operations for letters you see.

- To find interest rate (r), put your finger over r. Perform the operations for letters you see.

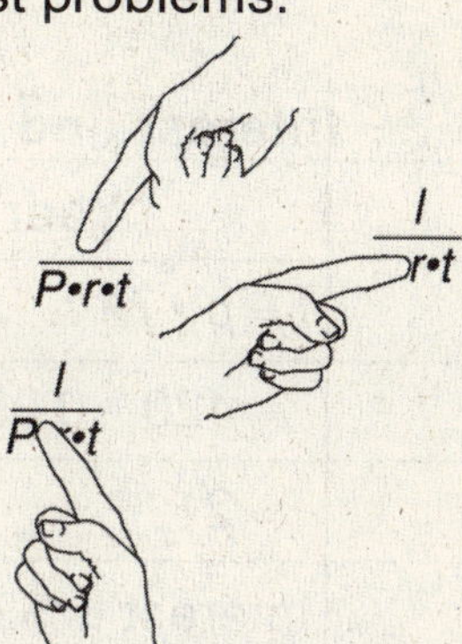

Find each missing value.

1. $p = \$400$, $r = 5\%$, $t = 3$ years

$I = p \cdot r \cdot t$

$I = \underline{\hspace{3cm}} \cdot \underline{\hspace{2cm}} \cdot \underline{\hspace{2cm}}$

$I = \underline{\hspace{3cm}}$

2. $p = \$15,000$, $r = 6\%$, $t = 2$ years

$I = p \cdot r \cdot t$

$I = \underline{\hspace{2cm}} \cdot \underline{\hspace{2cm}} \cdot \underline{\hspace{2cm}}$

$I = \underline{\hspace{3cm}}$

3. $I = \$350$, $r = 7\%$, $t = 2$ years

$p = \dfrac{I}{r \cdot t}$

$p = \dfrac{\underline{\hspace{2cm}}}{\underline{\hspace{1cm}} \cdot}$

$p = \underline{\hspace{3cm}}$

4. $I = \$168$, $p = \$1,400$, $t = 4$ years

$r = \dfrac{I}{p \cdot t}$

$r = \dfrac{\underline{\hspace{2cm}}}{\underline{\hspace{1cm}} \cdot}$

$r = \underline{\hspace{3cm}} = \underline{\hspace{2cm}}$

5. $I = \$57$, $p = \$380$, $t = 5$ years

$r = \underline{\hspace{3cm}}$

6. $p = \$4,800$, $r = 6\%$, $t = 2$ years

$I = \underline{\hspace{3cm}}$

7. $I = \$1,200$, $r = 4\%$, $t = 4$ years

$p = \underline{\hspace{3cm}}$

8. $p = \$750$, $r = 7\%$, $t = 3$ years

$I = \underline{\hspace{3cm}}$

Holt McDougal Mathematics

Challenge

Adding On

Simple interest is the amount of interest earned on the original principal. However, you can earn interest on the interest as well as on the principal. This is called **compound interest.**

Josef deposits $400 in a bank that pays 5% interest, compounded annually. Find the amount of interest and principal in Josef's account after 3 years.

Interest and Principal

Year 1	Year 2	Year 3
$I = p \cdot r \cdot t$	$I = p \cdot r \cdot t$	$I = p \cdot r \cdot t$
$= 400 \cdot 0.05 \cdot 1$	$= 420 \cdot 0.05 \cdot 1$	$= 441 \cdot 0.05 \cdot 1$
$= 20$	$= 21$	$= 22.05$
Interest is $20.	Interest is $21.	Interest is $22.05.
Principal is $420.	Principal is $441.	Principal is $463.05.

The amount of interest earned after 3 years is
$20 + $21 + $22.05 = $63.05.

The amount of principal plus interest after 3 years is
$400 + $63.05 = $463.05.

You can also find the total of principal and interest on $400 for 3 years at 5% by multiplying on a calculator $400 \cdot 1.05 \cdot 1.05 \cdot 1.05$. This shows the amount of interest and principal in Josef's bank account after 3 years, or $463.05.

Find the total amount of interest and principal if interest is compounded annually.

1. $500 for 2 years at 4%

____

2. $1,200 for 3 years at 4.5%

____

3. $300 for 4 years at 5.5%

____

4. $750 for 2 years at 5.5%

____

5. $98 for 3 years at 6.5%

____

6. $1,056 for 4 years at 5.25%

____

7. $520 for 5 years at 4.75%

____

8. $873 for 6 years at 5.2%

____

     Holt McDougal Mathematics

LESSON 6-7
Problem Solving
Simple Interest

Write the correct answer.

Use the graph to solve Exercises 1–3.

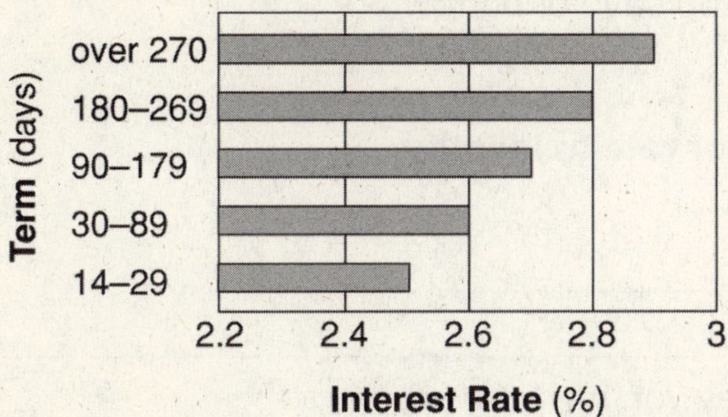

1. How much more interest would be earned on a $100,000 CD for 9 months than for 6 months?

   ____

2. A customer earned $3,262.50 interest on a 9-month CD. How much was the opening deposit?

   ____

3. Mrs. Wallace bought a $125,000 CD with a term of 3 years. How much will she earn in 3 years?

   ____

4. Until June 2002, the simple interest rate on Stafford loans to college students was 5.39% while the student was still in college. How much interest would a student pay on a $1,500 loan for 2 years?

5. Diego deposits $4,200 into a savings account that pays 5% simple interest. He decides not to touch the money until it doubles. How long will Diego have to keep the money in this account?

Choose the letter of the best answer.

6. Scott took out a 4-year car loan for $5,500. He paid back a total of $7,370. What interest rate did he pay for this loan?

   A 9.5%   C 8.5%

   B 9%   D 7.5%

7. How much interest would you earn if you were to deposit $575 for 3 months at 2.88% simple interest?

   F $4.14   H $41.40

   G $4.83   J $48.30

8. How long would you need to keep $775 in an account that pays 3% simple interest to earn $93 interest?

   A 4 years   C 4 months

   B 2 years   D 2 months

9. If you borrow $12,000 for 30 months at 6.5% simple interest, what is the total amount you will have to repay?

   F $12,065   H $13,950

   G $12,780   J $21,500

                    Holt McDougal Mathematics

Name ____ Date _________________ Class _______________

Reading Strategies
Focus on Vocabulary

Principal is the amount of money you save or borrow from a bank.

Interest is the amount of money the bank pays you for the use of your money, or the amount of money you pay the bank to borrow its money.

Rate is the amount of interest paid on money you save or borrow. Rates are usually given as percents.

For Exercises 1–3, write principal, interest, or rate to identify each situation.

1. You have $250 in a savings account.

   ____

2. The bank pays you 4% a year on the money you have saved.

   ____

3. The bank paid you $10 on your savings account last year.

   ____

If you wanted to borrow $3,000 for two years at a rate of 6%, you could use this formula to find the amount of interest you would pay:

$$\text{Interest} = \text{principal} \cdot \text{rate} \cdot \text{time}$$
$$I = p \cdot r \cdot t$$

$$I = \$3{,}000 \cdot 6\% \cdot 2$$
$$I = \$3{,}000 \cdot 0.06 \cdot 2 \quad \leftarrow \text{Change percent to decimal.}$$
$$I = \$360 \quad \leftarrow \text{Multiply.}$$

4. What formula is used to find the interest for this loan?

   ____

5. What is the decimal for 6%?

   ____

6. How do you find the amount of interest that will be paid on a loan?

   ____

   ____

Holt McDougal Mathematics

<table>
<tr><td>**LESSON**
6-7</td></tr>
</table>

Puzzles, Twisters & Teasers

Simply Interesting!

Use what you know about simple interest to complete the chart.
Then use your answers and the answer key to solve the riddle.

Principal	Interest Rate	Time	Simple Interest
$3,455	4%	S	$691
F	5.25%	2 years	$630
$16,500	E	24 months	$2,310
A	6%	3 years	$135
$625	3.5%	10 years	C

What do people in clock factories do all day?

They make _______ _______ _______ _______ _______ .
    $6,000 $750 $218.75   7%   5 years

      Holt McDougal Mathematics

Answers

LESSON 6-1

Practice A

1. $\dfrac{30}{100} = 30\%$ 2. $\dfrac{45}{100} = 45\%$

3. $\dfrac{67}{100} = 67\%$ 4. $\dfrac{1}{5}$

5. $\dfrac{8}{25}$ 6. $\dfrac{9}{10}$

7. $\dfrac{43}{100}$ 8. $\dfrac{12}{25}$

9. $\dfrac{29}{50}$ 10. $\dfrac{71}{100}$

11. $\dfrac{4}{5}$ 12. $\dfrac{7}{25}$

13. $\dfrac{17}{20}$ 14. $\dfrac{21}{50}$

15. $\dfrac{13}{25}$ 16. 0.19

17. 0.43   18. 0.07

19. 0.62   20. 0.23

21. 0.36   22. 0.59

23. 0.03   24. 0.725

25. 0.098   26. 0.0704

27. 0.1249

Practice B

1. 38%   2. 44%

3. 72%   4. $\dfrac{4}{25}$

5. $\dfrac{49}{100}$ 6. $\dfrac{1}{5}$

7. $\dfrac{3}{20}$ 8. $\dfrac{9}{50}$

9. $\dfrac{3}{5}$ 10. $\dfrac{7}{20}$

11. $\dfrac{23}{50}$ 12. $\dfrac{43}{50}$

13. $\dfrac{79}{100}$ 14. $\dfrac{14}{25}$

15. $\dfrac{9}{20}$ 16. 0.33

17. 0.57   18. 0.46

19. 0.06   20. 0.047

21. 0.132   22. 0.758

23. 0.04   24. 0.0116

25. 0.2705   26. 0.9301

27. 0.079

Practice C

1. 28%   2. 40%

3. 56%   4. $\dfrac{29}{100}$; 0.29

5. $\dfrac{7}{8}$; 0.875   6. $\dfrac{3}{40}$; 0.075

7. $\dfrac{16}{25}$; 0.64   8. $\dfrac{5}{8}$; 0.625

9. $\dfrac{17}{50}$; 0.34   10. $\dfrac{9}{40}$; 0.225

11. $\dfrac{1}{8}$; 0.125   12. $\dfrac{9}{250}$; 0.036

13. $\dfrac{59}{1000}$; 0.059   14. $\dfrac{9}{125}$; 0.072

15. $\dfrac{41}{200}$; 0.205   16. >

17. =   18. <

19. >   20. <

21. >   22. =

23. >   24. <

Review for Mastery

1. 38; $\dfrac{19}{50}$ 2. 20; $\dfrac{1}{5}$

3. 70; $\dfrac{7}{10}$ 4. 16; $\dfrac{4}{25}$

   Holt McDougal Mathematics

5. $36; \dfrac{9}{25}$

6. $8; \dfrac{2}{25}$

7. $\dfrac{3}{20}$

8. $\dfrac{53}{100}$

9. $\dfrac{6}{25}$

10. $\dfrac{17}{100}$

11. 0.58

12. 0.93

13. 0.15

14. 0.09

15. 0.26

16. 0.02

17. 0.8

18. 0.01

19. 0.235

20. 0.096

21. 0.407

22. 0.0703

Challenge

Event A	**Event B**
$P(A) = 45\%$	$P(B) = \dfrac{11}{20}$
$P(A) = 70\%$	$P(B) = \dfrac{3}{10}$
$P(A) = 42\%$	$P(B) = \dfrac{29}{50}$
$P(A) = 17.5\%$	$P(B) = \dfrac{33}{40}$
$P(A) = 65\%$	$P(B) = 0.35$
$P(A) = 37.5\%$	$P(B) = \dfrac{5}{8}$
$P(A) = 48\%$	$P(B) = \dfrac{13}{25}$

F I N N I S H

Problem Solving

1. $\dfrac{17}{25}$; 0.68

2. 73%

3. $\dfrac{23}{50}$; 0.46

4. 41%; 0.41

5. C

6. F

7. D

8. J

Reading Strategies

1. per hundred

2. as fractions and as decimals

3.

4. 0.40

5. $\dfrac{40}{100}$; $\dfrac{2}{5}$

6. 50%

Puzzles, Twisters & Teasers

1. $\dfrac{13}{20}$

2. $\dfrac{1}{5}$

3. $\dfrac{18}{25}$

4. $\dfrac{63}{100}$

5. $\dfrac{3}{50}$

6. $\dfrac{6}{25}$

7. 0.0362

8. 0.6

9. 0.52

10. 0.063

11. 0.45

12. 0.362

E V A P O R A T E D   M I L K

LESSON 6-2

Practice A

1. >

2. =

3. >

4. <

Holt McDougal Mathematics

5. <

6. >

7. 34%

8. 8%

9. 19%

10. 64%

11. 56.1%

12. 4.9%

13. 10.5%

14. 90%

15. 25%

16. 14%

17. 93.75%

18. 45%

19. 87.5%

20. 22.5%

21. 22%

22. 80%

23. Possible answer: Since the denominator 25 is a factor of 100, mental math is a good choice; 68%.

Practice B

1. 17%

2. 56%

3. 4%

4. 70%

5. 2.5%

6. 80.3%

7. 30%

8. 7.2%

9. 32.5%

10. 60%

11. 15%

12. 41.7%

13. 31.25%

14. 3.75%

15. 83.3%

16. 76%

17. 22%, 0.3, $\dfrac{19}{40}$

18. $\dfrac{1}{11}$, $\dfrac{2}{19}$, 11%

19. 5.8%, $\dfrac{5}{8}$, 0.675

20. −0.35, −21%, 0.051

21. $\dfrac{1}{9}$, 27%, 0.351

22. −$\dfrac{4}{5}$, 0.08, 0.8

23. Possible answer: Since the denominator 25 is a factor of 100, mental math is a good choice; 24%.

Practice C

1. 4%

2. 97.5%

3. 3.1%

4. 40.9%

5. 37.8%

6. 40%

7. 52.4%

8. 6.7%

9. 16.25%

10. 77.8%

11. 85%

12. 62.5%

13. 97.5%

14. 36.4%

15. 44%

16. 18%

17. −0.41, −12%, $\dfrac{3}{25}$

18. $\dfrac{3}{7}$, 43%, $\dfrac{4}{9}$

19. −$\dfrac{11}{12}$, 94.5%, 0.98

20. 83%, $\dfrac{5}{6}$, $\dfrac{17}{20}$

21. 0.41, 43.5%, $\dfrac{9}{2}$

22. 0.55%, 48%, $\dfrac{13}{24}$

23. Possible answer: Since $\dfrac{30}{80} = \dfrac{3}{8}$, and 3 ÷ 8 divides evenly, pencil and paper is a good choice; 37.5%.

Review for Mastery

1. 34%

2. 6%

3. 93%

4. 57%

5. 80%

6. 73.4%

7. 8.2%

8. 22.5%

9. 60.4%

10. 9%

11. 51.8%

12. 3.9%

13. 30%

14. 4%

15. 35%

16. 20

17. 12.5%

18. 12%

19. 75%

20. 57.5%

21. 55%

22. 86%

23. 96%

24. 87.5%

Holt McDougal Mathematics

Challenge

Possible path shown.

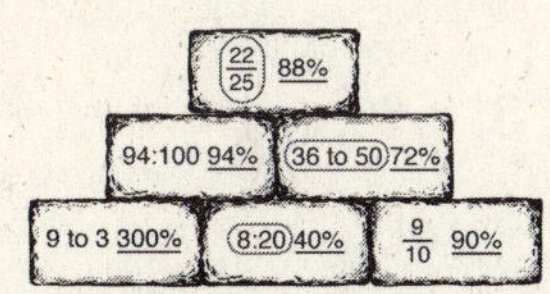

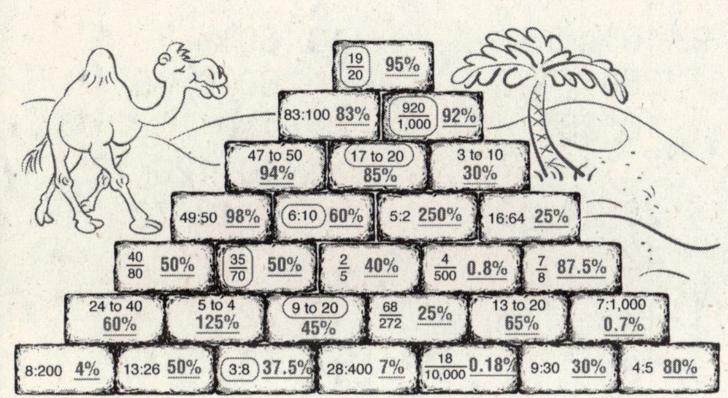

Problem Solving

1. 13.7%   2. 55%

3. mashed potatoes, 2%

4. 68.75%   5. D

6. J   7. D

8. J

Reading Strategies

1. $\dfrac{7}{10}$ 2. 70%

3. 0.45   4. 45%

5. Possible answer: Change the labels on the first bar so that the bar shows a total of 200, and each square represents 10 units. Then find the percent that is equivalent to 110 units.

Puzzles, Twisters & Teasers

1. $\dfrac{3}{4}$ 75%

2. $\dfrac{5}{9}$ $55.\overline{5}$%

3. $\dfrac{13}{16}$ 81.25%

4. 0.84   84%

5. 0.55   55%

6. 0.075   7.5%

7. $\dfrac{7}{40}$ 17.5%

8. $\dfrac{9}{20}$ 45%

9. 0.045   4.5%

10. 0.825   82.5%

S U B M A R I N E S

LESSON 6-3

Practice A

1. C   2. G

3. A   4. H

5. B   6. J

7. D

Answers may vary. Possible answers given.

8. 20   9. 33

10. 60   11. 30

12. 20   13. 4

14. 6   15. 120

Answers may vary. Possible answers given.

16. 10   17. 2

18. 6   19. 35

20. 4   21. 30

22. 2   23. 27

Practice B

Answers will vary. Possible answers are given.

1. 16   2. 14

3. 82   4. 7

Holt McDougal Mathematics

5. 50   6. 9
7. 30   8. 45
9. 150   10. 4
11. 45   12. 270
13. about $8

Answers may vary. Possible answers are given.

14. 15   15. 32
16. 6   17. 12
18. 2   19. 54
20. 5   21. 84
22. 30   23. 42
24. 84   25. 4
26. $36,400

Practice C

Answers may vary. Possible answers are given.

1. 13   2. 12
3. 55   4. 4
5. 24   6. 13
7. 330   8. 90

Answers may vary. Possible answers are given.

9. 24   10. 54
11. 16   12. 9
13. 5   14. 7
15. 81   16. 220

Answers will vary. Possible answers are given.

17. $2.30; $43.70   18. $1.00; $17.00
19. $1.20; $22.30   20. $2.00; $36.00
21. $5.00; $90.00   22. $0.70; $12.00

Review for Mastery

Possible answers:

1. 20%; $\frac{1}{5}$ 2. 60%; $\frac{3}{5}$

   50;   300;

$\frac{1}{5}$; 50; 10 $\frac{3}{5}$; 300; 180

10   180

3. 5%; $\frac{1}{20}$ 4. 50%; $\frac{1}{2}$

   40;   530;

$\frac{1}{20}$; 40; 2 $\frac{1}{2}$; 530; 265

2   265

5. 9   6. 150
7. 70   8. 48

Possible answers:

9. 10%; 5   10. 60%
   280   60; 6; 10
   10; 28   120
   5; 14   10; 120; 12
   28; 14; 42   6; 12; 72
   42   72

11. 12   12. 72
13. 16   14. 120

Challenge

Estimates may vary.

Problem Solving

Answers will vary. Possible answers are given.

1. about 25 visitors

Holt McDougal Mathematics

2. Midwest region

3. about 140 visitors

4. about 850 shoppers

5. D
6. H

7. A
8. H

Reading Strategies

1. $\frac{1}{3}$
2. 10%

3. $\frac{1}{4}$
4. estimate

5. exact answer
6. estimate

7. $33\frac{1}{3}$%
8. $\frac{1}{2}$

9. 10%

Puzzles, Twisters & Teasers

S: 117

T: 54

A: 30

E: 31

C: 10

H: 20

Y: 26

W: 6

P: 4

L: 41

O: 5

R: 8

THEY WERE ALL
COPYCATS

LESSON 6-4

Practice A

1. 24
2. 12

3. 6
4. 1.9

5. 60
6. 129

7. 45
8. 216

9. 72
10. 128

11. 81
12. 27

13. 15
14. 16

15. 3
16. 75

17. 8
18. 128

19. 6
20. 161

21. 435
22. 70

23. 125
24. 9

25. 116ºF

Practice B

1. 14
2. 11

3. 7.5
4. 162

5. 60
6. 198

7. 7
8. 570

9. 187.5
10. 108.1

11. 319.8
12. 0.32

13. 495
14. 70

15. 96
16. 18

17. 13.6
18. 23.1

19. 0.77
20. 475

21. 22.5
22. 405

23. 16.2
24. 29

25. 0.95
26. 33

27. 213.2
28. 30

29. about 0.5 inch

Practice C

1. 87.8
2. 13.9

3. 9
4. 0.7

5. 96
6. 415.4

7. 430.9
8. 430.9

9. 15.5
10. 17.7

11. 10.7
12. 5.3

13. 1.3
14. 111.0

15. 0.2
16. 714.1

17. 137.5
18. 44.85

19. 5.7
20. 688

21. 126.28
22. 531.25

23. 171
24. 198

Holt McDougal Mathematics

25. about 2,047 radio stations

Review for Mastery

1. a. x
   b. 90
   c. 40
   d. $\dfrac{x}{90} = \dfrac{40}{100}$
   e. 3,600
   f. 36

2. a. x
   b. 520
   c. 85
   d. $\dfrac{x}{520} = \dfrac{85}{100}$
   e. 44,200
   f. 442

3. 24
4. 141
5. 90
6. 84
7. 26
8. 19
9. 46
10. 33

Challenge

1. 24.7
2. 76.76
3. 23.8
4. 58.22
5. 62.92
6. 43.2
7. 17.35
8. 77.76
9. 57.66
10. 88.32
11. 128.4
12. 126.66
13. 144
14. 200

Problem Solving

1. more than 1.8 billion
2. Asia
3. about 1.2 billion
4. about 1.2 billion
5. B
6. G
7. D
8. F

Reading Strategies

1. 0.35
2. by moving the decimal point two places to the left
3. multiplying the decimal times the number
4. 0.10
5. 6
6. 12
7. 18
8. 24

9. Possible answer: The answers increased by multiplies of 6.

Puzzles, Twisters & Teasers

1. Y
2. O
3. U
4. R
5. S
6. H
7. A
8. D
9. O
10. W

Y O U R   S H A D O W

LESSON 6-5

Practice A

1. D
2. H
3. C
4. G
5. D
6. F
7. C
8. G
9. 60%
10. 10%
11. 80%
12. 60%
13. 22
14. 200
15. 120
16. 75
17. 140
18. 500
18. 4%

Practice B

1. 125
2. 48
3. 30%
4. 12.5%
5. 625
6. 200
7. 40%
8. 8%
9. 50
10. 25
11. $6\dfrac{2}{3}\%$
12. 60
13. 140
14. 8%
15. 12.5%
16. 40%
17. 1,100
18. 70
19. 6.5%
20. 4.25%

Practice C

1. 50
2. 25
3. 200
4. 11.1%

   Holt McDougal Mathematics

5. 13.3%   6. 20
7. 20%   8. 33.3
9. 350   10. 75%
11. 240%   12. 16.9%
13. 275   14. 85.2
15. 87.5   16. 40
17. 16.7%   18. 23.1%
19. 76.5   20. 30%
21. 8.75%

Review for Mastery

1. a. 14   2. a. 16
   b. 25   b. x
   c. x c. 80
   d. $\frac{14}{25}$ d. 16; 80
   e. 25; 1,400   e. 80; 1,600
   f. 56   f. 20
   g. 56%   g. 20
3. 55%   4. 40
5. 300%   6. 140

Challenge

1. 89   2. 14
3. 100   4. 25
5. 96   6. 71
7. 100   8. 93
9. 83   10. 60
11. 33   12. 7

Problem Solving

1. 66%   2. 5.5%
3. 96%   4. 15% off
5. D   6. G
7. B   8. H

Reading Strategies

1. •   2. =
3. n 4. $6 = 10\% \cdot n$
5. 27 is what percent of 30?
6. $n \cdot 40 = 30$

Puzzles, Twisters & Teasers

L: 25
R: 90
A: 140
O: $33.\overline{3}$
V: 40
C: $66.\overline{6}$
I: 12.5
K: 20
T: 60
F: 50
S: 200
E: 300
A ROCK FESTIVAL

LESSON 6-6

Practice A

1. $25 - 17 = 8$; 25; $8 \div 25 = 32\%$
2. $36 - 24 = 12$; 24; $12 \div 24 = 50\%$
3. $50 - 40 = 10$; 50; $10 \div 50 = 20\%$
4. $56 - 40 = 16$; 40; $16 \div 40 = 40\%$
5. 75%   6. 40%
7. 300%   8. 75%
9. 33.3%   10. 90%
11. 60%   12. 11.1%
13. $13.35; $75.65
14. 40%; percent decrease
15. $23.98; $35.97
16. 10%; percent increase

Practice B

1. 45%   2. 25%
3. 75%   4. 300%
5. 150%   6. 16.7%
7. 61.9%   8. 121.5%
9. 71.4%   10. 90.5%
11. 52.6%   12. 19.4%
13. 69.5%   14. 366.7%

    Holt McDougal Mathematics

15. 3.2%   16. 87.5%

17. $12.87; $26.13

18. $43.97; $35.98

19. $12,750   20. $40.80

Practice C

1. 62%   2. 20

3. 17.5%   4. 166.7%

5. $100   6. $2.88

7. 1.1   8. 36.8%

9. 67.1%   10. 83.3

11. 27.6%   12. 68.4%

13. 5   14. 132.8%

15. 46.2%   16. 40.5

17. $704.88; $5,169.12

18. $225.25   19. 231.8%

Review for Mastery

1. 14
   8
   $\frac{14}{8}$; 175

2. 9
   90
   $\frac{9}{90}$; 10

3. 75
   125
   $\frac{75}{125}$; 60

4. 340
   400
   $\frac{340}{400}$; 85

5. 25%   6. 95%

7. 80%   8. 40%

9. 200%   10. 5%

Challenge

1. 289; 332   2. 189; 204

3. $4.70; $3.76   4. 419

5. $1.27

6. Fun Electronics, since the game will cost $40.96; it will cost $41.36 at Gamers.

Problem Solving

1. about 7.6 million cars

2. 31.1% increase

3. 3% increase   4. 137 pounds

5. 36.2% increase

6. 15.8% decrease

7. C   8. G

9. C   10. J

Reading Strategies

1. $50   2. decrease

3. in the denominator (or bottom part) of the fraction

4. 25   5. 20

6. $\frac{20}{25}$

Puzzles, Twisters & Teasers

PATIENTS

LESSON 6-7

Practice A

1. $1,000; 0.05; 2   2. $600; 0.04; 3
   $100 $72

3. $330; 0.03; 1   4. $270; 0.05; 3
   $11,000 $1,800

5. $600; $2,500; 4   6. $108; $900; 3
   6%   4%

7. $75   8. $840

9. $3,750   10. $5,800

11. 5%   12. 7%

13. 4 years   14. $2,000

15. 6.5%   16. $5,850

    Holt McDougal Mathematics

Practice B

1. $225
2. $480
3. $375
4. $1,200
5. 4%
6. 6%
7. $15.75
8. $768
9. $1,875
10. $9,600
11. 5%
12. 7%
13. $29.25
14. $900
15. 6%
16. 3.5%
17. 5 years
18. $5,975
19. 5.5%
20. 14.3 years

Practice C

1. $210
2. 7%
3. 7 years
4. $9,400
5. 4.5%
6. $67.50
7. $307.50
8. 12 years
9. $4,000
10. 2 years
11. 2.5%
12. $330
13. 2.5 years
14. 9.5%
15. 4.5%
16. about 18 years
17. $4,700
18. $9,975

Review for Mastery

1. 400; 0.05; 3
   $60
2. $15,000; 0.06; 2
   $1,800
3. $\dfrac{350}{0.07 \cdot 2}$
   $2,500
4. $\dfrac{168}{1,400 \cdot 4}$
   0.03; 3%
5. 3%
6. $576
7. $7,500
8. $157.50

Challenge

1. $540.80
2. $1,369.40
3. $371.65
4. $834.77
5. $118.38
6. $1,295.84
7. $655.80
8. $1,183.34

Problem Solving

1. $775
2. $150,000 deposit
3. $10,875
4. $161.70
5. 20 years
6. C
7. F
8. A
9. H

Reading Strategies

1. principal
2. rate
3. interest
4. $I = p \cdot r \cdot t$
5. 0.06
6. Multiply the principal by the rate by the length of time on the loan.

Puzzles, Twisters & Teasers

S: 5 years
F: $60,000
E: 7%
A: $750
C: $218.75
F A C E S

Holt McDougal Mathematics